The Renaissance, The Reformation and The Rise of Science

The Renaissance, The Reformation and The Rise of Science

Harold P. Nebelsick

T&T CLARK
EDINBURGH

T&T CLARK
59 GEORGE STREET
EDINBURGH EH2 2LQ
SCOTLAND

All rights reserved. No part of this publication may be reproduced, stored in a retrieval system, or transmitted, in any form or by any means, electronic, mechanical, photocopying, recording or otherwise, without the prior permission of T&T Clark.

First Published 1992

ISBN 0 567 09605 1

British Library Cataloguing-in-Publication Data
A catalogue record for this book is available from the British Library

Typeset by Trinity Typesetting, Edinburgh
Printed and bound in Great Britain by Billing & Sons Ltd, Worcester

Table of Contents

Preface	v
Introduction	vii

Chapter 1: The Christian Critique of Aristotle — 1

The Arab Contribution	1
The Bearers and Developers of Greek Learning	2
The Arab Achievement	5
The Christian Milieu	10
Philoponos, the Pioneer	11
Grosseteste, Scientist at Oxford	18
Roger Bacon's Anti-Aristotelian Science	30
The Inductive Method of Duns Scotus	46
An Assessment of Aristotelianism	49
Ockham's Empirical Logic	53
The Fall of Aristotle	62

Chapter 2: The Renaissance Mind — 79

The Roots of the Renaissance	79
The Place of Plato	91
Plotinus, Neo-Platonist	97
The Systematic Theology of Proclus	105
Hermeticism's Cosmic Cult	112
Ficino, Renaissance Platonist	118
The "New Science"	129

Chapter 3: The Reformation and the Rise of Science — 148

Biblical Faith and The Rise of Science	149
Creatio Ex Nihilo	149
Covenant Relationship	161
Meaningful History	169
The Move Toward Natural Science	189
Francis Bacon's England	189
The Protestant "Scientist" Francis Bacon and The Hermetic Imagination	196
Francis Bacon's Way of Science	209
Epilogue	234

PREFACE

Harold Paul Nebelsick was Professor of Doctrinal Theology at Louisville Theological Seminary and a Member of the Center of Theological Inquiry in Princeton. When he died on Easter Sunday, 1989, this book was not yet ready for publication. Realizing that he had indeed written a work of considerable value, those who knew him and of his accomplishment chose to take it upon themselves to put the finishing touches to his manuscript. The desire to share his ideas with the public grasped us all. The following words are for those who would like to know more about this undertaking.

On Easter Sunday the manuscript which was to become this book lay in pieces. The lion's share was complete. However, there were sections which could not be included for a variety of reasons. Among Professor Nebelsick's notes there is a substantial essay devoted to Giordano Bruno, entitled " Giordano Bruno: Renaissance Magus". We decided to exclude it, in view of its content, for it was evidently about to develop into a book on its own. It is far too lengthy to include here, and its content diverges from the main scope of the book. It is possible that it might be published separately. There is also a very short section on Johannes Kepler, entitled "Johannes Kepler: The New Astronomy". This was intended to complement that on Francis Bacon, but was not fully developed. The decision was made to remove it altogether.

All editorial decisions were based upon the principle of remaining as true as possible to the phraseology and intention of the author. Although this was not always a simple task, there were no major hurdles, for the manuscript of what appears here was left in a sufficiently advanced stage. We were faced with only one sizable problem: there were two chapters on Francis Bacon. They were slightly different in content, but the same in argumentation. In addition it was evident that Professor Nebelsick intended to develop further his critical assessment of Bacon's thought. The decision was made to compose a single chapter on Bacon using the two extant copies. All incomplete commentary was removed. In this way we felt we would include all that he had written on Francis Bacon for this book without adulterating his argument. Had Professor Nebelsick lived longer he would have certainly pruned it severely, cutting out unnecessary repetitions and reiterations, but this is something which wwe did not feel free to undertake ourselves.

We believe that in this work Harold P. Nebelsick has contributed something substantial to our understanding of the relationship of theology to scientific culture. He breaks through our prejudices concerning both theology and science and places their relationship into perspective in new and startling ways. Few have grasped the impact of theological and biblical thought on scientific culture with the depth presented here. This book is a significant contribution to the theological renaissance we are now experiencing and toward which he was so deeply committed.

Special thanks are due Mr William Witherspoon of St Louis and to Louisville Presbyterian Theological Seminary, to the Center of Theological Inquiry in Princeton, to Professors W. Jim Neidhardt and Boris Kuharetz of New Jersey Institute of Technology, Dr Hugh Old, and to Professor Samuel and Mrs Eileen Moffett of Princeton, to Professor Dietrich Ritschl of Heidelberg, and Professor C. F. von Weizsäcker of Starnberg, for their generous support. Without the advice of Professors Jim Neidhardt and Thomas F. Torrance the preparation and publication of this book would not have been possible. Finally our deepest appreciation goes out to Melissa Nebelsick who participated in this project from its earliest stages.

Paul Matheny and Mary Nebelsick,
Wilson, North Carolina
October 1901.

INTRODUCTION

As scholars like Professors Günter Howe (1908-1968), C. F. von Weizsäcker (1912-), and Thomas F. Torrance (1913-) have emphasized, science in the West developed when the elements of Greek science, brought to the West by the Arabs, were commingled with the Christian understanding of God and his relationship to the world expressed by the biblical doctrines of the Protestant Reformation. It is true that a certain elementary knowledge of Hellenistic science had been kept alive in the West by the works of Greek and Latin Church Fathers as well as by the writings of Pliny (c.23-79) and Boethius (c.479 - c.524). The accumulation of knowledge throughout the Middle Ages made it possible for scientists to understand the spherical nature of the earth and the arrangement and motion of the heavenly bodies. By the eleventh century cosmology was again understood at the level which it had reached in the days of Plato (c.427 - c.347 B.C.) some fifteen hundred years before.[1]

Meanwhile in the East, especially in Syria, in centers like Harran and Bagdad, the Arabs learned to appreciate Greek science, particularly that of Aristotle (384-322 B.C.) and Ptolemy (c.90 - c.168). In the latter half of the twelfth century they brought the texts of Euclid (fl. c. 300 B.C.), Aristotle and Ptolemy, along with their commentaries upon them and the records of their own astronomical observations, to the West by way of Sicily and the Iberian peninsula. Thus, they laid the foundations for a quite unprecedented revival of learning in Europe. The ancient knowledge inspired the western mind to raise classical thought to the highest level of authority and to follow its precepts in theology, philosophy and natural science. This revival of learning led to the Renaissance, the Protestant Reformation in the sixteenth-century, and eventually the scientific revolution of the seventeenth century.

Nevertheless, the rediscovery of ancient "science", particularly that of Aristotle and Ptolemy, inspirational as it was to the revival of learning in Oxford and Paris,[2] did not lead to the immediate birth of modern science. This was due, in part, to theological and concomitant philosophical reasons. In the West during the late Middle Ages, as in Greece in the age of Pericles (c.495 - 429 B.C.), theological and philosophical thought served both to encourage a rudimentary development of science and to hold that development in check. In this regard the West followed the pattern of ancient Greece. Except for the pre-Socratic Ionian philosophers who attempted to conceive the material world as consisting of material parts, Greek thought concerning the world (science) and God (theology) were so commingled that science was prevented from developing beyond its initial stages. In the West, the Greek theologically-determined philosophy of Plato, Aristotle, and, an amalgam of both, neo-Platonism continued to compromise conceptions of reality in such a way that the investigation of nature had difficulty in gaining a foothold.

The Middle Ages was dominated by the neo-Platonism of Augustine (354-430) which so devalued nature that it was neither studied nor considered worth studying.[3] When, in the thirteenth century, Aristotelian "scientific treatises" along with Aristotelian philosophy began to influence the western mind, initially through the efforts of Averroës [Ibn Rushd], (1126-1198) in Spain, Robert Grosseteste (c.1170-1253) at Oxford, Albert the Great (c.1200-1280), and his pupil, Thomas Aquinas (1225-1274) in Paris, interest in nature was revived, at least to a certain extent. However, as far as its fundamental concepts were concerned, Aristotle's philosophy proved to obstruct the development of science. Like neo-Platonism Platonism it continued to represent reality in such a way as to discourage the investigation of nature for its own sake. Like Plato who viewed nature as unreal, a shadow of the ideal, Aristotle considered matter to be of secondary importance. It was an "accident" of the *substance* of reality. Added to that, Aristotle's "metaphysics" so circumscribed his "physics" with the contradictory characteristics of rigidity and capriciousness that investigation of nature was discouraged. The rigidity resulted from its claim that first and final causes both determined reality and interpenetrated it in such a way that all solutions could be gained from present principles by the simple application of deductive logic. At the same time and equally important, Aristotle's understanding of reality in organismic terms subjected nature to the unpredictable and capricious intervention of living and divine principles.

However, lest we lay too much blame on theology and philosophy for holding back the development of science, as indeed they did, it may be well to remind ourselves that science is inevitably inter-related with one kind of theologico-philosophical conceptuality or another. With the early Babylonians and Greeks, it was the theological prescription of the divine, eternal and perfectly circular movements of the heavens that first attracted their interest in the stars in order to fathom the ways of God. Eventually the tracing of their movements and the calculation of their orbits gave birth to astronomy as a "science".

The appreciation of Aristotle in the late Middle Ages and early Renaissance followed the same pattern. In the West Aristotle was appreciated as much for his philosophical and theological thought as for his ideas about nature. Just as in the ancient world, philosophy and theology led people to look to the heavens but in part prevented them from seeing what was there, so in the West, philosophical and theological thought at one and the same time enhanced interest in science but prevented its proper development.

With regard to astronomy, for instance, the ancient Greek theology designated the heavenly bodies as divine and required that their movements, which were thought to reflect divinity, be perfectly circular, regular and immutable. This combined with Aristotle's concept of the unmoved mover who caused the planets to move whether by friction between the heavenly spheres or by the intervention of "star-spirits" complicated the picture of the heavens. Whereas the perfectly regular and circular orbital movements of the heavenly bodies depicted the heavens as an assemblage of orderly movements, the star spirits, who looked to the mover of all things for direction but moved the heavenly bodies according to their own tergiversating wills both explained and introduced a dimension of non-predictability and capriciousness into the heavenly system. Similarly, in the Aristotelian-influenced West of the late Middle Ages and early Renaissance, this same theology combined divine cause with the pattern of divine circularity. Thus, science was prevented from developing beyond the stages it had already reached with Ptolemy well over a thousand years before. Wherever Aristotle reigned, natural science both began and was held at bay.

I will argue, first, that at the time Greek science was transmitted to the West, the theologico-philosophical concepts that were part and parcel of it so defined God and described reality that, while Aristotelian "science" intrigued the West as a description of nature, it also served to hinder the development of modern science. Second, theological conceptualities of the bizarre combination of neo-Platonic, Hermetic, and Renaissance philosophy and those of the sixteenth-century Reformation served to enhance the development of science. Hence, instead of showing that the interaction between theology and natural science was either inevitably salutary or always harmful, the evidence would seem to indicate that at times it was counter-productive and at other times beneficial.

Stanley Jaki (1924-) indicates one side of this influence when he shows that late Greek theology had debilitating effects upon the development of physics from its early beginnings in Greece to the rise of modern science in the West. The religious motivation that helped decide the course of physics for almost two thousand years was epitomized by Aristotle. Following Socrates' and Plato's lead, he made physics serve the "living and divine principles" that were believed to rule the cosmos.[4]

Oddly enough, however, it was just such a philosophy of living and divine principles revived by a new Platonism that was eventually to give impetus to the development of modern science. In its neo-Platonic and Hermetic forms the concept of the divine as interpenetrating, rejuvenating, and recreating all things served to help break the hold that the Peripatetics (Aristotelians) of the thirteenth to fifteenth centuries, with their rationalistic deductivism, had on the middle Renaissance mind.

In the seventeenth century, the esoteric ideas of the "new science" based on neo-Platonic ideas inspired the Renaissance imagination to re-investigate the universe. When this imagination was tamed by the theology of the Reformation of the sixteenth century, science could begin in earnest. By following biblical doctrine, according to which God was understood as God and the world as the world, Reformation thought secularized the "living and divine principles" whether of Plato, neo-Platonism or Hermeticism, and compromised the authority of Aristotelian rationalistic deductivism. As a result imagination and order were able to meet and dialogue with one another. Out of that conversation modern natural science was born. Modern natural science, I shall argue, depends upon faith, imagination, and a sense of order—these three—and all are equally important.

As far as faith in relation to natural science is concerned, we will move beyond the rather rudimentary observation that all science requires faith. Science and indeed all thought would be quite impossible without faith defined as trust in that which is beyond rational demonstration. Testimonies in this regard abound. Max Planck (1858-1947), for example, said, "Science demands also the believing spirit", and "over the entrance to the gates of the temple of science are written the words: Ye must have faith."[5] Werner Heisenberg (1901-1976) called faith "the mainspring of scientific endeavor" and correlated faith with the work of science saying, "I believe in order that I may act: I act in order that I may understand".[6] Albert Einstein (1879-1955) referred to faith as basic to science in a number of respects: belief in the intelligibility of nature, belief in the existence of a universe structured according to definite models and revealing orderly arrangement, and belief that nature is conceivable in the simplest mathematical ideas.[7] In 1931 in an address, "Clerk Maxwell's Influence on the Evolution of the Idea of Physical Reality", Einstein said, "Belief in an external world independent of the perceiving subject is the basis of all natural science".[8] Ten years later, when assessing the influence of science, religion and philosophy, he explained that the source of feeling imbuing the scientist with the aspiration for truth and understanding "springs from the sphere of religion . . . I cannot conceive of a genuine scientist without that profound faith. Science without religion is lame, religion without science is blind".[9] In this sense, of course, "faith" as trust in the existence, faith in the stability and the knowability of the world is a part of life itself. Augustine's *credo ut intelligam*, "I believe in order to know", is a maxim which applies to all knowledge whatsoever as well as to theological knowledge in particular.[10]

The question therefore, is not "whether faith or not?" but "which faith?" Belief systems are both basic to and exert a deep influence on our attitudes to nature as well as to God. Christianity has played a major constitutional role since the year 315 when Constantine (c.288-337), for better or worse, designated Christianity as a lawful religion of the empire. Hence, the question we must ask with regard to the complex mixture which makes up western culture is not whether or not theology is involved in the development of natural science but, "What is the nature of the theology that has influenced that development and what effect has it had upon it?"

As Alfred North Whitehead (1861-1947) and Herbert Butterfield (1900-) both have testified, it would seem quite legitimate to compare the import of the rise of science, which began to change the face of the western world in the seventeenth century, with that of the rise of Christianity in the first.[11] Butterfield compares the development of science to Christianity in the Middle Ages which dominated life completely and came "to preside over everything else, percolating into every corner of life and thought".[12] Because of science, the face of the earth and the activities of men were to alter more in a century than they had previously done in a thousand years".[13] He then goes on to argue for the beneficial influence of the Christian faith upon the rise of modern science in the West.

It cannot be said, of course, that the Church greeted its competitor gladly or shared its position over life and culture willingly or intentionally. The imprisonment of Roger Bacon (c.1214 - c.1294) in 1278 at the hands of his fellow Franciscans, Martin Luther's (1483-1546) lack of appreciation of Nicholas Copernicus' (1473-1543) thought, the burning of Giordano Bruno (1548-1600), a Copernican Hermeticist, and the preliminary process against Galileo Galilei (1564-1642) in 1616, when Copernicus' writings were placed on the *Index of Forbidden Books*, along with the Inquisition's condemnation of him and his Copernican concepts of the universe in 1633, would seem sufficient evidence of the reluctance of the ecclesiastical authorities to accept the rise of modern science even in its rather rudimentary development.[14]

The decrees against the teaching of Descartes at the universities of Leiden and Utrecht inspired by the hyper-Calvinist, Gisbert Voetius (1589-1693) and the ban on Cartesianism, issued by the Archbishop of Paris, as late as 1680 in the face of the philosopher Pierre-Sylvain Régis' (1632-1707) use of the philosophy of Descartes as the introduction to scientific thinking, show that Protestant and Roman Catholic could agree in their condemnation of the kind of thinking that was basic to modern science. The general outcry of ecclesiastics against Charles Darwin's (1809-1882) *The Origin of Species* in 1859, led particularly by the Anglican Bishop Samuel Wilberforce (1805-1873), an outcry that is continued even today by the so-called "Creationists," gave and continues to give the widespread impression that the Church was and is the entrenched enemy of science as such.[15]

The fact that the 1969 General Assembly of the Presbyterian Church in the United States (Southern) rejected a previous General Assembly statement that had pitted the Bible against the Darwinian theory of evolution was a long time in coming. Since there is a movement abroad in the Roman Catholic Church to reconsider its verdicts against Bruno and Galileo and since Orthodox, Roman Catholic and Protestants now meet in conference after conference on theology and science, it certainly appears that we have entered a new day of rapprochement between science and the Christian faith. In the early seventeenth century, however, freedom from ecclesiastical influence was, in fact, often a prerequisite for the development of science.

In this regard it is of more than passing interest to note that the University of Padua in northern Italy (where such men as Andreas Vesalius [1514-1564], Matteo Colombo [c.1516-1559], Hieronymus Fabricius [1537-1619], and William Harvey [1578-1657] as well as Copernicus, Galileo and Bruno, all studied) owed its freedom of thought—an absolute necessity to the development of science—to the fact that in 1404 Padua had fallen under the political power of Venice. Until the Reformation, Venice was the most staunch anti-clerical state in Europe.[16] The University of Göttingen, which in the nineteenth and the first half of the twentieth centuries was and still is eminently noted for its achievements in the natural sciences, offers another pertinent example of an institution whose freedom from clerical control was productive of scientific thought. Such scientists as Carl Friedrich Gauss (1777-1855), [Georg Friedrich] Bernhard Riemann (1826-1866), Max Planck (1858-1947), Max Born (1882-1970), Werner Heisenberg (1901-1976), C. F. von Weizsäcker (1912-), and Manfred Eigen (1927-), perhaps owe more than they may themselves have realized to the policy of Göttingen University that specifically curtailed the authority of its theological faculty. From its founding in 1734, in contrast to the practice of other universities of Germany, Göttingen forbade the theological faculty the right of censorship over the faculty of natural science.

The burning of Michael Servetus (1511-1553) in sixteenth-century Geneva, albeit for theological rather than scientific reasons, and the seventeenth century decrees of the universities of Utrecht and Leiden against Cartesianism, inspired by Gisbert Voetius (1589-1676) and his followers in the Netherlands, are indication enough that freedom of thought could not be taken for granted in Protestant lands. It is, therefore, hardly surprising that in a day of reconciliation between theology and science, we often find a formal anti-ecclesiastical tendency among scientists. Günter Howe's youthful experience some fifty years ago tallies with that which too often marks the teaching of natural science in our time.

Even for the generation of my teachers who helped me to pass my High School final examinations in the year 1943, the "Age of Invention and Discovery," the great age of heroes, and the "Copernican Revolution" of modern times, was due to the liberation of scientific thinking from feudal and clerical guardianship, a process which came to full self-consciousness during the Enlightenment.[17]

Nevertheless, although it is true, as Thomas S. Kuhn (1922-) points out, that "only the civilizations which descend from Hellenic Greece have possessed more than the most rudimentary science",[18] it is also true, and hardly coincidental, that it was specifically in the Christian culture of Europe, and that of northern Europe in particular, that the scientific-technological revolution began to come to full flower. Kuhn designates sociological factors as the reason for this development, claiming that "the bulk of scientific knowledge is a product of Europe in the last four centuries", and adds the point that "no other place and time has supported the very special communities from which scientific productivity comes".[19] However, he gives no indication as to the basic structure of the societies out of which these special communities arose, nor does he trace the sources of the frame of mind nor the motivations that gave impetus to those who devoted themselves to reconceiving and reconstructing their world.

If C. F. von Weizsäcker is correct in saying, "Modern science would not perhaps have been possible without Christianity",[20] then we have an apparently contradictory situation. Although the church establishment, for the most part, was opposed to the development of natural science or, at best, indifferent to it, nevertheless science as we know it came into being within the culture of Europe which was influenced and shaped by the Christian faith. We understand this somewhat ambiguous state of affairs when we recognize that although the development of science took place, often in opposition to the expressed intention of the church, it nevertheless arose on the basis of the very message which the church proclaimed, the faith it propagated, and the doctrines it taught.

As so often happens, the support given by the faith to the growth of science illustrates the fact that the truth proclaimed has a life of its own. It thus moves beyond the control of the particular institutions or persons promulgating it. It is new wine which inevitably bursts and pours forth from the old wine-skins thought to be quite adequate by the viniculturists who trusted their product to them. If so, the fact that natural science developed most vigorously in "Christian" Europe, and particularly in Protestant-dominated northern Europe, may be considered more than a matter of fortuitousness. The development, I shall argue, is due to a coincidence which resulted from a combination of Greek philosophical (including scientific) thinking, on the one hand, and Christian theological thought, on the other.

Our account will begin with a critique of Aristotelian thought that was the dominant force in the investigation of nature from its inception during the fourth century B.C. in Greece until the renaissance of learning in the West in the early twelfth century. When the Arabs transferred the writings of Aristotle, their commentaries upon his writings, and the writings of Euclid and Ptolemy to the West in the twelfth century, a revival of interest in classical learning arose. Residual interest in Platonic thought characteristic of Augustinian theology, was revived and intensified. The Scriptures of the Old and New Testaments were re-evaluated and re-examined.

In the middle Renaissance a renewed interest in Plato, neo-Platonism and Hermeticism led to a Platonic Renaissance and the "new science". The "new science" had less to do with the investigation of nature than with the explication of imaginative ideas about the mystical aspects of the divine Soul which supposedly interpenetrated all things. Nevertheless, when it was combined with the kind of cultic conceptualities and practices that were closely aligned with the Pseudo-Dionysian piety propagated by Thomas Aquinas, the "new science" developed into an alternative system of thought to that of Aristotle.

At the same time, the renewed interest in the ancient Scriptures of the Old and New Testaments led to a challenge of Aristotelian-informed medieval theology by the Protestant Reformers. On the basis of biblical doctrines they set Aristotle aside and repudiated the ecclesiastical institution that continued to be dependent on the Aristotelian deductive system. Their proclamation of God, who was transcendent over nature rather than being intermixed with it, served to demythologize nature. Thus, nature could be put into "proper perspective" so that it could be investigated and known.

Hermetic neo-Platonism which had fired the Renaissance mind of Marsilio Ficino (1433-1499) and may well have aided Nicholas Copernicus in considering the heliocentric system as a viable alternative to Ptolemy's geocentricity, continued along side by side and sometimes in conjunction with the thought of the Reformation. Under its influence Aristotelian thought began to lose its authority. Aristotle as a rationalistic deductivist fell from the pinnacle of dominance, and he continued to deaden the sensibilities of those who persisted in seeking to re-establish his former triumph. However, both the "new science" of the Renaissance and Reformation thought were very much alive.

Odd as it may seem, the commingling of the "new science" and Reformed thought proved to be the right mixture for the birth of science. The vivid imagination of the Renaissance mind had revived the logic of induction and a biblical appreciation of humankind as observer and manipulator of creation while Reformed thought renewed hope in history and tamed nature for investigation. This was the combination that made modern science possible.

Francis Bacon (1561-1626) provides an excellent example of the way it came about. Perhaps he more than any other can be called its midwife.

NOTES

1. J. L. E. Dreyer, *A History of Astronomy from Thales to Kepler*, rev. ed. (New York: Dover, 1953), p. 227. Cf. H. P. Nebelsick, "Science Encounters the Christian Faith," *Circles of God* (Edinburgh: Scottish Academic Press, 1985), pp. 88 ff.
2. A. C. Crombie, *Augustine to Galileo*, 2 vols. in 1 (London: Heinemann, 1979), II, 76 ff.
3. This thought, according to Günter Howe, is from Karl Jaspers who uses it to characterize Greek thought as a whole. Günter Howe, *Mensch und Physik* (Witten: Eckert Verlag,1963), p. 35.
4. Stanley Jaki, *The Relevance of Physics* (Chicago: University of Chicago, 1970), p. 412.
5. Max Planck, *Where is Science Going?* (New York: Norton, 1932), p. 214.
6. Werner Heisenberg, "A Scientist's Case for the Classics", *Harpers* (May, 1958), 29.
7. Albert Einstein, *The World as I See It* (New York: Covici, 1934), "On the Method of Theoretical Physics", pp. 36-39.
8. *Ibid.*, "Clerk Maxwell's Influence on the Evolution of the Idea of Physical Reality", p. 60.
9. Albert Einstein, *Out of My Later Years* (Secaucus, N.J.: Citadel, 1974), p. 26.
10. The fiduciary aspect of knowledge is being stressed most particularly in our day by Michael Polanyi, *Personal Knowledge* (Chicago: University of Chicago, 1958), pp. 264 ff., and *The Tacit Dimension* (New York: Doubleday, 1967), p. 61.
11. Herbert Butterfield, *The Origins of Modern Science 1300-1800* (London: Bell and Sons, 1973), p. 190. A. N. Whitehead, *Science and the Modern World* (New York: Free Press, 1969), p. 2.
12. Butterfield, *Origins of Modern Science*, p. 179.
13. *Ibid.*
14. In all probability Bruno was burned more for his heretical theological ideas than for his heretical Copernican cosmology. However, since the records to the trial have been lost and in view of Galileo's first trial in 1616 which, like Bruno's, featured Cardinal Robert Bellarmine (1542-1621), it is difficult to judge whether or not Bruno's heterodox cosmology was a part of the offense or not. Cf., Stanley L. Jaki, *The Road of Science and the Ways to God* (Edinburgh: Scottish Academic Press, 1978), p. 45; Giordano Bruno, *The Ash Wednesday Supper* (*La cena de le ceneri*), tr. Edward A. Gosselin and Lawrence S. Lerner, "Introduction," pp. 11-34.
15. Roland M. Frye, *Is God a Creationist? The Religious Case Against Creation-Science* (New York: Scribner's, 1983).

16 Butterfield, *Origins of Modern Science*, p. 48.
17 Günter Howe, *Gott und die Technik* (Hamburg: Furche-Verlag, 1971), p. 33.
18 Thomas S. Kuhn, *The Structure of Scientific Revolutions*, Second Edition Enlarged (Chicago: University of Chicago, 1970), p. 168.
19 *Ibid.*
20 C. F. von Weizsäcker, *The Relevance of Science* (London: Collins, 1964), p. 181.

Chapter 1

THE CHRISTIAN CRITIQUE OF ARISTOTLE

The Arab Contribution
Any attempt to trace what is now often referred to as the "intellectual history" of the West requires at certain junctures that we take a step backward in order to proceed forward in a somewhat orderly fashion. In the Introduction we began by stating that the three major forces shaping the western thought which led to the rise of modern science were Platonism, Aristotelianism and the teachings of the Christian faith. From their inceptions in Periclean Greece, Platonic and Aristotelian thought vied and alternated with one another in capturing the loyalties of the western mind. Down deep, however, at the basis of their epistemological structures there is more agreement than disagreement between these two main carriers of Greek thought even though the differences are significant.

Platonic thought tends to reflect a world substantiated by the unity and mystery of transcendent Being of which the material world is a mere reflection, a world of shadow. In Aristotle, too, the real, the result of first and formal causes, is hidden from observation. The real *substance* shows itself under the guise of material accident. The order of logic follows that of being (ontology). To know the world one argues deductively from *known* first principles and orders the phenomenal world accordingly.

Platonism traveled the road from the early centuries of our era to the Renaissance largely under the guise of neo-Platonism undergirded and transported by the theology of Augustine. As we will develop in the next chapter, the system, if indeed it can be called a "system", was inaugurated by Plotinus (c.205 - c.270), systematized by Proclus (c.410-485), put into an even more esoteric but religiously seductive form of Hermeticism and introduced into the Renaissance in its Latin versions by Marsilio Ficino's translations.

Greek science did not fall from heaven in the West any more than the concepts on which it was built suddenly appeared among the early philosophers of ancient Greece. Analogous to the way we now know that the rudiments of science were transferred from the Babylonians to Greece by the early Ionian philosophers—Thales (c.636 - c.546 B.C.), Anaximander (c.611 - c.547 B.C.), and Anaximenes (c.550 - c.480)—Greek science was brought to the West by the Arabs.[1]

The Bearers and Developers of Greek Learning
It is somewhat ironic, perhaps, but not without a certain degree of poetic justice, that the Arabs are not only responsible for sustaining modern western industrial civilization with their oil, but in a real sense, are responsible for the transfer to Christian European culture of the intellectual basis on which our oil-dependent civilization developed in the first place. In view of the fact that the Arabs and their civilization are, by and large, little understood and appreciated in the West, it may help to facilitate understanding if we show our dependence on Arab civilization and learning by delineating the route by which the Arabs brought Greek learning to the western world.

Alexandria, the city of Ptolemy, which had been established by Alexander the Great (356-323 B.C.) in 332 B.C,. fell to the Arab forces in 646 A.D. Subsequently the impetus toward learning in general and scientific learning in particular which had previously been moved from Athens to Alexandria during the reign of Alexander, was relocated from the mouth of the Nile to the banks of the Tigris. In Bagdad about 800 Caliph Harun al-Rashid (c. 764-809) had Ptolemy's *He mathematike syntaxis* translated into Arabic. The Arabs called it the *Almagest* as do we. Another step was taken by Caliph Al-Mamun (c.786-833) who built a "house of wisdom" in Bagdad in 830 which included a grand observatory, a museum and a library. The museum and library emulated those at Alexandria which suffered final destruction with the razing of the pagan temples under Theodosius I (347-395) in 391. It was at this institution in Baghdad that Albumazar (c.805–885) observed the skies, made his notations, and cast horoscopes. There, too, Abd-al-Rahman al-Sufi (903-986) made the observations that afforded the data he needed to revise Ptolemy's star catalogue. He was followed by Abulwefa (c.940 - c.997) who preceded Tycho Brahe (1546-1601) by 500 years in making and noting continuous planetary observations. Even more impressive was Ibn Yunis' (c.950-1009) discovery of the method of determining time by altitudes. He noted his observations in the "Hakemite Tables" in which he set down the values of astronomical constants with extreme accuracy. His fixings of the solar eclipses of 977 and 978 were made with such precision that they were satisfactorily used for notations to designate the acceleration and deceleration of the moon in its orbit.

Arab astronomy continued in the East in the twelfth and thirteenth centuries, even as the Arabs made their way through North Africa, crossed the Mediterranean and settled in Sicily and the Iberian peninsula from where they began to influence western thought in earnest. In Persia, Hulagu Khan (1217-1265) established an observatory at Maragha. Equipment included a mural quadrant twelve feet in radius as well as altitude and azimuth instruments of the type employed by Tycho Brahe four centuries later. In the middle of the thirteenth century Nasir ed-Din (1201-1274), who had become the director of the observatory at Maragha, drew up the "Ilkhanic Tables" which determined the annual rate of precession of the earth within one minute of current value. Observations were continued a century and a half later by Ulugh Beg (1394-1449) who founded his observatory at Samarkand and made it a center of astronomy that compared favorably with that which Tycho Brahe established on the island of Hveen in the latter part of the sixteenth century. From observations he made at Samarkand, Ulugh Beg both redetermined nearly all the stars listed by Ptolemy and published astronomical tables whose accuracy was respected as standard for the next two hundred years.

It is hardly surprising, therefore, that when the Arabs settled the Iberian peninsula and made Cordova and Toledo centers of learning, they took their interest in astronomy with them. There they continued to study the heavens and to emphasize the locations and positions of the stars. As early as 1080, Alzarkali (c.1029 - c.1087) edited the "Toledan Tables". In 1252 Alfonso X of Castile (1221-1284) was responsible for the publication of the "Alfonsine Tables" which retained his name.

Certainly of equal importance to their primary observations was the trasmission to of the Arabic translations of Greek scientific and philosophical writings to the West. Among these was Ptolemy's *Almagest* translated by Gerard of Cremona (c.1114-1187) in 1175 who spent most of his life in Toledo at a college of translators established by Archbishop Raymond (d. 1151). With this achievement the most prodigious example of Greek science became available to western scholars. The *Almagest*, which in Arabic means "the greatest", was the name given by the Arabs to Ptolemy's *Syntaxis*.[2] It must, by all counts, be considered one of the premier scientific treatises of all time. The book is a paradigm of clarity, thoroughness and integrity. It includes or refers to almost all we know of the development of Greek astronomy.

Ptolemy began the work with a short section, "Hypothesis on the Planets", in which he explained the geocentric universe that he had inherited from Aristotle and his successors, modified the epicycles of Apollonius of Perga (c.262-c.190 B.C.), the eccentrics of Hipparchus (fl. 146-127 B.C.) and added equants, an invention of his own.[3] He then explained the mathematics and spherical trigonometry which he used in formulating his theorems, detailed the angle of the horizons as they appeared at different latitudes on earth, explained the movements of each of the bodies and their different appearances, and provided the whole system with relevant notations and tables.

Following Aristotelian physics, Ptolemy explained that the geometry of the heavens combined the characteristics of circularity and sphericity. Because they considered the heavenly bodies to be divine, to each was attributed its own particular will. In addition, Ptolemy continued to maintain Aristotle's differentiation between mathematics, on the one hand, and physics, on the other. Mathematics was the abstraction from the physical pattern of things while physics was the actual representation of the heavenly movements themselves.[4] He also continued to adhere to Aristotle's division between the divine, eternal, ethereal, unchanging, incorruptible and rational heavenly supra-lunar spheres which enunciated themselves through perfect circularity, and the temporal, material, mutable, corruptible, irrational sub-lunar earthly sphere characterized by rectilinear motion.[5] So thorough, complete and persuasive was this most felicitous application of Aristotelian theology, metaphysics, and physics to astronomy that it was to dominate the way the world was seen and understood for the next fifteen hundred years. Not even Copernicus was able to break with its basic geometric tenets based on the perfect regularity and circularity of the movements of the heavenly bodies.[6]

The Arab Achievement
The achievement of Arab civilization, of course, was not confined to the science of astronomy or to the transference of Hellenistic learning to the West. In the Middle Ages the Arabs dominated the Mediterranean and stretched their influence around it in a crescent from Anatolia to India, back through Arabia and Africa and finally north to the Iberian peninsula. They developed an admirable level of culture and built the fabulously wealthy cities of Damascus and Bagdad at a time when western Europe was relatively uncivilized. They both appropriated, appreciated and preserved Greek classical learning and built upon it. For generations they led the world not only in astronomy but in medicine as well. Their development of art, culture, and learning was such that they produced the Koran, wrote highly sophisticated poetry and excelled in mathematics.

The Arab mathematical ability and the sense of symmetry it inspires expressed itself not only in astronomy but in art and architecture. Their application of geometry to architecture is especially ingenious. On the Iberian peninsula, where the Arab Ommayads developed a highly sophisticated culture during their residence, this masterful integration of mathematics with art and architecture is still in evidence.

At Cordoba, the city from which the philosopher Averroës [Ibn Rushd] spread classical learning to the rest of Europe, the eighth century mosque with its two tiers of arches possesses a symmetry and grace that was not to be matched until the descendents of the barbarian Goths pointed the Arab arch in their "Gothic" cathedrals. Disfigured as it is by just such a Gothic cathedral set down in its midst, the mosque at Cordoba remains a lasting monument to beauty and harmony brought about by setting the harmonies of mathematics into stone.

The Alhambra Palace at Granada, surrounded by geometrically-designed, fountain-filled gardens with water running freely in rectilinear channels and with row upon row of exotic plantings bordering symmetrically designed courts, is likewise mathematics made manifest. Inside the palace the domed cedar-wood ceilings, horseshoe-arched windows, sculptured stucco ornamentation, and finely moulded pierced plaster-work in polygonal patterns cover the walls and ceilings. These along with the geometric designs of the ceramic tiles of the baths continue to testify to beauty brought about by the application of geometry to art and architecture.

Investigations of Arab civilization and learning such as those of F. E. Peters and Sayyed Hossein Nasr, make a strong case for the independent development of Arab culture from its Hellenistic base.[7] The evidence indicates that from the time the Arabs fell heir to Alexandrian Aristotelianism, Arab culture developed out of the vigorous intellectual ferment that arose in conjunction with its encounter with Hellenistic thought. There were, of course, the orthodox Hanbalites who demanded complete subservience to the literal teachings of the Koran. There were also the somewhat less conservative Ash'arites who, though professing absolute subjection to the will of Allah, used philosophical categories to interpret it. More important for Arab *falsafah* (philosophy) and the development of natural science, however, were the Mu'tazilites who, though guided by the Koran, took an equally strong stance on the validity of free will, sense perception, intuition, and speculation.

When Islam began to assimilate Hellenistic learning and Arab philosophy and modern science began to develop, Arab culture underwent the same kind of painful, and to a degree fruitful, enlightenment that the West experienced at the time of the Renaissance. Arguments and counter-arguments were put forth; schools were formed; textbooks were written; learning proceeded; and traditions arose that had their time of influence and receded again when their intellectual and/or political bases were eroded.

In Islam, as in Christianity later, there was a deep and ever recurring inner debate as to the legitimacy and place of science and philosophy alongside of and as integrated with theology. Aristotelian thought in particular was assimilated by Islamic theology long before Thomas Aquinas used it as a basis for his theological synthesis in the West and it continued through and beyond Thomas' lifetime. A century after Thomas, Ibn Khaldun (1332-1406) sums up the movement which had begun to make itself felt in Islam as early as the middle of the ninth century.

> Recent speculative theologians, then, confused the problems of theology with those of philosophy, because the investigations of theology and philosophy go in the same direction, and the subject and problems of theology are similar to the subject and problems of metaphysics. Theology and metaphysics, thus, in a way come to one and the same discipline. The recent theologians, then, change the order in which the philosophers had treated the problems of physics and metaphysics. They merged the two sciences in one and the same discipline . . . .The science of speculative theology thus merged with the problems of philosophy, and theological works were filled with the latter.[8]

If so, the Arabs may have assimilated Greek thought and Aristotle, in particular, not too wisely but too well. If such is the case with Islam in the East as it was later with Christianity in the West, Aristotle may well have led the Arabs into science, but he could only lead them so far and no farther. The coincidence between Aristotelian rationalism and that of the Koran was so strong that the *falsafah* of such thinkers as Avicenna [Ibn Sina] (980-1037), Alghazali [Algazel] (1059-1111), and Averroës [Ibn Rushd] (1126-1198) was not able to overcome the Islamic belief in Allah's supreme control of the world, and its events. On the one hand this control was rationalistic and deterministic; on the other it was unfathomable and even capricious. This fatal combination of fate and fickleness endemic to Aristotle hindered the development of Arab natural science just as it had Greek science. In spite of the high level of Arab learning, their artistic, and technical ability as well as the subtleties of their philosophical debates, they had come to a dead end.

Although they were not able to develop a highly sophisticated science themselves, they made an enormous contribution to science. Indeed they made the development of science possible in the West by bringing Greek learning and philosophy, especially the works of Aristotle, to Europe, and by writing commentaries upon them. It was primarily the writings of Averroës that, when transferred to Paris in the thirteenth century, enabled Thomas Aquinas to combine the philosophy of Aristotle with traditional Augustinianism and to construct the "medieval synthesis". This was to determine the direction of western theology until the sixteenth century, and influences it still.

The impact of Aristotle's writings on both philosophy and science that led to the appreciation of Greek learning was supplemented by Euclid. Euclid's measurements of reality with the theorems and formulae of his plane geometry astounded people then and continue to elicit our admiration for their order, comprehensiveness and simplicity. Ptolemy took Aristotle's geocentric system of the world, added to it the cycles, epicycles and eccentrics of the later Pythagoreans, plus equants of his own making, arranged it all according to Euclid's format and notated it by means of his geometric formulae.

When the writings first of neo-Platonic Hermeticism and

then of Plato himself were translated, the minds that changed the medieval and Renaissance worlds were astounded by both the breadth and the profundity of ancient thought. At one and the same time it reopened investigation into the physical world, a world almost forgotten and it taught thought to soar into the upper echelons of reality and beyond it to a realm designed and designated by unbridled speculation. Thus, the re-appropriation and re-appreciation of classical learning led to the revival of science as well as to the renewed study of classical thought in general. The results were the Renaissance in the thirteenth century, the Protestant Reformation in the sixteenth century, and eventually the rise of modern science in the seventeenth.

Arab sympathy for Aristotlian philosophy was, of course, no coincidence. Allah exists in regal majesty over the world and he rules it with supreme divine power and inexorable will. Like Aristotle's first and final cause, Allah's will pervades the whole universe and operates with the resolute consistency of absolute necessity. According to the Koran, Allah is the almighty creator whose justice is to be feared (Sura XVI, vs. 1 ff.). Allah, the "Merciful One" (Sura IV, vs. 44) dispenses his mercy intentionally and prescriptively: "Verily Allah will cause to err when he pleases and will direct whom he pleases" (Sura XXXV, vs. 9). As interpreted by the orthodox Hanbalite theologians and even by the less conservative Ash'arites, this meant that all activities of creation were so predestined by God, that reality was determined from beginning to end. Allah's inexorable will left little or no freedom for innovative or constructive thought. This kind of consistent rationalism was constructive of a closed system that did not inspire a great deal of enthusiasm for the kind of investigation, speculation and innovation that is necessary for modern scientific procedure.

The theological factor which Joseph Needham found blocking scientific development among the Chinese had similar effects. It was not that there was no order in nature for the Chinese, but rather that it was not an order ordained by a rational personal being. Hence, there was no conviction that rational and personal beings would be able to articulate in lesser earthly languages the divine code of laws which the Creator had decreed before time.[9]

As with the conservative Arab Hanbalites, so among the Chinese, predetermined fixed order was the rule. Thus, in spite of the fact that the Chinese had progressed technologically to the point where they could pass on to western Europe the magnetic compass in the thirteenth century, the rocket in the fourteenth and printing in the fifteenth, experimental science, as we know it, seems not to have developed among the Chinese any more than it did among the Arabs. Stephen Toulmin makes the point that the Chinese emperors were the guardians of this order, "heterodox speculations about mathematical astronomies served no useful purpose. So, instead of welcoming new conceptions as a mark of intellectual vigor, they regarded them with suspicion".[10] The development of experimental science that is utterly dependent on both belief in the stability of order and the freedom to depart from a particular conception of order was stayed off.

The Christian Milieu
Modern science developed in the seventeenth century in post-Reformation western Europe rather than among the Chinese, the Greeks or the Arabs, all of whom possessed the seeds of that development. The question that arises therefore, is, "What were the ingredients that made up the amalgam of post-Reformation western European culture, in general, and that of Protestant Europe, in particular that can be said to have looked upon science with favour?" "Why was it," asks C. F. von Weizsäcker, "that it was western Europe that provided the seed-bed in which modern science could root and grow?"

> Islamic culture was for long centuries of the Middle Ages philosophically, scientifically and in the realm of medicine superior to European culture. Why did modern science come into being in Europe and not, for example, in the Arab world? The tradition of Greek philosophy and science as well as the tradition of the faith as portrayed by the biblical account of creation was common to both cultures. Where does the difference come from?[11]

The difference, I would argue, following especially C. F. von Weizsäcker and T. F. Torrance, is the Judeo-Christian faith. It is at least arguable that it was the teachings or doctrines of the biblically-oriented faith that provoked the European mind to open itself to the reality of nature as *nature* and to the world as *world*.[12] Three teachings, in particular, would be candidates of most importance for having inspired the cast of mind, the intellectual climate, and the appropriate atmosphere necessary for the development of science. It was the extended application of the Greek practices of observation, classification, and mathematical quantification of things to nature which led to the development of modern science and technology. The first teaching that needs to be mentioned in this regard is the biblical *doctrine of creation*. The second, closely bound to the first, is the *doctrine of redemption* that undergirds the biblical understanding of history. The third, related to both the first and the second, is the *doctrine of work*.

It must not be forgotten, that although science and technology arose under the influence of the church, in many instances it arose in direct opposition to ecclesiastical authority. Nonetheless, as stated in the introduction to this volume, a case can be made for the position that the faith which the church propagated was an indispensable factor in the conception of reality in relation to which science was to develop. This very science was often opposed by church authorities with such vigour that outrageous consequences ensued.

Philoponos, the Pioneer
The Christian church had inherited its doctrine of the Creator from the Jewish Scriptures. Jahweh (God) was, at one and the same time, Creator and Lord. "In the beginning God created the heavens and the earth." God signified his lordship both by creating all that came to be and by being Lord over all he created. According to the Genesis account, the earth that was without form and the darkness that was upon the face of the deep, all signify forboding chaos. God brings forth order by creating light, dividing

light from darkness, the heavens from the earth, the water from the land, and day from night. He made the heavenly bodies (divine bodies to other ancient cultures) and ordered the sun to rule the day, the moon the night. Fish, birds, animals and finally humankind are all created and ordered according to the Word of God who created the world and ordered that which he created. The creation, later interpreted in the Christian Scriptures as "bringing into existence that which did not exist" and in the history of the early church as *creatio ex nihilo* (creation out of nothing) exhibited God's Lordship over the whole of creation. The doctrine was to become of utmost importance for the development of science and technology.

A thousand years before the Reformers of the sixteenth century re-emphasized the doctrine of *creatio ex nihilo* in theology[13] the Christian philosopher John Philoponos (c.490-566) of Alexandria had already used the doctrine to challenge Aristotle. Philoponos, a pupil of Ammonios (fl. 5th cent. A.D.) and one of the last important philosophers of the Alexandrian school, is beginning to receive much deserved attention in our day, not the least because he was perhaps the first to recognize the implications of the Christian doctrine of creation for science.[14] For Philoponos, the concept of God as Creator, who created the world *ex nihilo*, was hardly compatible with either the implicit identification of God with the deified uppermost heavens implicit in Aristotle's thinking or with the consequent Aristotelian insistence on a differentiation in substance between that of heaven and that of earth.[15]

Wolfgang Wieland has argued well that Philoponos, a physicist and theologian, predates Thomas Aquinas as the first Christian thinker of note to use the thought structures of Aristotle's philosophy to explicate the Christian faith.[16] It is even more important to realize that Philoponos predates the rise of modern science by renouncing a great deal of the physics and thought about the world propagated by Plato and Aristotle as early as the sixth century. Jaki makes the point that Philoponos understood "for the first time the purifying effects of theological tenets in physical reasoning".[17]

Following his perception that God was responsible for the creation of the whole universe, Philoponos was convinced that the cosmos as a whole was composed of the *same kind of matter* and was subject to the *same laws*. Hence, in direct opposition to prevailing thought, he both rejected the dichotomy between the *finite earthly* and the *infinite eternal* heavenly realms and recognized the importance of earthly reality. Further, especially in contrast to the neo-Platonism of his day, Philoponos insisted that nature could not be understood as the finite representation of infinite reality but as real in itself. To understand reality one must make deductions based on observation. In contrast especially to Aristotle, he maintained that reality could not be apprehended by making deductions from *known principle*.

Thus, whereas in Aristotelian cosmology the planets were described as moving in perfect circles around a common point, observation dictated to Philoponos that the orbits deviated from concentricity. "'The stars each have their own specific movements along their spheres and around their own centers, not homocentric with the universe'".[18] The idea did not originate with Philoponos. Both Hipparchus' second century B.C. eccentrics and Ptolemy's employment of equants served the same purpose. The re-emphasis of the individuality of the movements of the heavenly bodies, which Plato had pointed out prior to Aristotle, namely that some stars display clockwise and others anti-clockwise movement, questioned Aristotle's concept of heavenly harmony at its roots.[19]

In addition Philoponos openly opposed Aristotle's statement in *De caelo* that the stars neither were made of fire nor did they move in fire. Interestingly enough, Philoponos first quotes the Apostle Paul to authorize his claim, "'One star differeth from another star in glory'". He then goes on to explain that, like everything else in the sub-lunar sphere, the colour and brightness of stars depend on differences in composition of matter as well as upon their form, size, movement and periods of revolution.[20] Contrary to the theories of both Plato and Aristotle that the heavenly bodies were composed of immaterial, divine essence, Philoponos claimed, "What is visible is in principle also tangible, and tangible things possess tangible qualities—hardness, softness, smoothness, roughness, dryness and humidity, as well as heat and cold which contain all the others".[21] Thus, in a return to the pre-Socratics, who had compared the heavenly bodies to earthly stones, Philoponos insisted that, generally speaking, there is nothing in the heavenly bodies which is not to be found in terrestrial bodies.[22]

Equally important for physics in general was Philoponos' theory of impetus which later became the basis for the theory of the inertia of bodies. Again, contrary to Aristotle, for whom the eternal movement of the heavenly bodies continued only by benefit of the constant impetus imparted to them by the "prime mover" Philoponos theorized that once the heavenly bodies were set in motion, God no longer had to intervene to keep them moving. In this way, he not only rejected Aristotle's theory which asserted that the continual movement of the heavens evidenced divine power at work, his theory of impetus antedated Galileo's mechanics by a thousand years.[23]

We now have evidence that Galileo supported his theory of the finite nature of the heavens on the basis of Philoponos.[24] It is more than a coincidence; rather a matter of design, that Galileo gave the name "Simplicius" to the Aristotelian-Ptolemaic protagonist in his fictitious debate on the Copernican system. It was the philosopher Simplicius (fl. c. 530) who had pleaded Aristotle's case against Philoponos in the sixth century. Because, according to Galileo, Simplicius lost the debate to Philoponos, he presents in his *Dialogue Concerning the Two Chief World Systems* "Simplicius", as again the loser, this time against "Filippo Salviati", representing Galileo himself.

Hence, it would seem that the more we learn, the more idols of modern science fall. As the precedent for the Copernican theory is to be found in the heliocentric system of Aristarchus of Samos, so the precedent for Galilean mechanics is to be found in Philoponos' theory of impetus. Walter Böhm, in fact, traces the influence of Philoponos' theory through such men as Avicenna [Ibn Sina] (980-1037), Avempace [Ibn Bajjah] (d. c.1138), Averroës [Ibn Rushd] (1126-1198), Roger Bacon (c.1214 - c.1294), John Duns Scotus (c.1266-1308), William of Ockham (c.1284 - c.1349), Jean Buridan (c.1300 - c.1358), Nicole Oresme (c.1320-1382), Nicholas of Cusa (c.1401-1464), Leonardo da Vinci (1452-1519), Giordano Bruno (1548-1600), and Galileo's forerunner, Giovanni Benedetti (1530-1590), to Galileo himself (1564-1642). It was Galileo who finally took advantage of Philoponos' theory in developing his "new mechanics".[25]

Recent research into the importance of Philoponos serves to document the beginnings of the rather circuitous route by which Greek learning reached the West. In the fifth and sixth centuries Alexandria was the center of learning at which the youth of wealthy families from all over the Near East, many of whom were Christian, congregated. At the School of Philosophy headed by Ammonios they studied law, medicine, mathematics and philosophy. Böhm speculates that it was on behalf of these Christian students that Ammonios, who was not a Christian, asked the Christian Philoponos to join him. The fact that Ammonios shared Philoponos' view that the world was created *ex nihilo* may or may not be coincidental. Another important factor may have been that the Christian catechetical school in Alexandria had long since taken heathen (*i.e.*, Greek) philosophy into its course of instruction. A reciprocal relationship or a mutual influence may well have existed between the School of Philosophy and the catechetical school.

In the sixth century, the School of Philosophy became Christian. In 641 the Arabs conquered Alexandria but the School of Philosophy seems to have survived until 718 when it was moved to Antioch. There, under aegis of the Antiochian patriarchs in the seventh century, a renaissance of Greek learning took place in the Syriac language. About 835 the School moved again, this time from Antioch to Harran. Lastly, about 850, it moved to Bagdad where it gradually became dominated by Islamic thought. Its heritage of Greek learning, especially the writings of Aristotle, was taken West when the Arabs conquered and occupied North Africa, Sicily and the Iberian Peninsula from the eighth to the fifteenth centuries.[26]

So understood, Philoponos not only takes precedence in breaking with Aristotle and the principles behind the science of dynamics, but stands at the threshold between Hellenistic learning and the Arab intellectual development. After moving through the East, Hellenistic learning came to the West where Aristotelian thought took hold. It was then compromised and finally broken by a combination of Platonic, neo-Platonic, neo-Pythagorean and Protestant Reformation thought.

Philoponos presents us with a case that, from the point of view of the development of science, is as tragic as that of Aristarchus of Samos (c.320 - c.250 B.C.). Advanced as was Aristarchus, who proposed the heliocentric universe some 1500 years before Copernicus, or perhaps just because he was so advanced, he did not carry the day. So, too, the "realism" of Philoponos was lost almost at its inception. So strong was the hold of Aristotle on the "scientific mind" and so contrary to both intuition and immediate observation, that Philoponos' theory seemed never to have been considered seriously.

From the fifth century onward and throughout the Middle Ages the dominant intellectual force in the West was that of Augustine. Unfortunately, as far as the development of natural science was concerned, Augustine never considered nature significant enough in and of itself to be considered of prime importance nor was knowledge of it considered of ultimate value. Augustine's neo-Platonism, like neo-Platonism in general, rests on Plato's concept of creation wherein time and earthly reality are but faint counterparts of primary, eternal and heavenly patterns. This world is but a shadow of the real.

As we have seen, Augustinian neo-Platonism with its implicit "doceticism" in regard to nature, like the neo-Platonism of Plotinus and Proclus and the combination of neo-Platonism and neo-Pythagorianism seen in Hermeticism, at best only *imagines* reality. Here one might apply Karl Jaspers' (1883-1969) statement about the Greeks to Augustine and the neo-Platonists in general, "The material side of the world was neither knowable nor worth knowing".[27] This would seem to challenge A. C. Crombie's statement that a rational inquiry into the nature of things is implicit in Augustine's theology through which western Christendom learned "to value the natural world as sacramental and symbolic of spiritual truths".[28]

It was exactly this evaluation of nature as having symbolic "sacramental" value which detracted from its intrinsic value. So long as matter was thought to persist primarily as a manifestation of *divine, eternal reality*, it had significance in and only insofar as the divine and eternal were reflected by it or through it. As Howe has said, to follow Augustinianism was to be in danger of pushing creation "on the rim of the theological field of vision".[29] Nature, understood in symbolic terms, manifested and signified the ways of God but had no ways of its own which were of real significance.

Some progress, was made in the direction of science when Thomas Aquinas identified Aristotle's rational "unmoved Mover" as the God of the Christian faith. It is here that we find the validity of Whitehead's contention that the rationality of God and nature as derived from Aristotle and incorporated by Thomas Aquinas into medieval theology was a necessary pre-requisite to the development of modern scientific theory.[30] To some degree, although not altogether, as we shall see with particular reference to natural order, we may also agree with Whitehead's statement that "faith in the possibility of science, generated antecedently to the development of modern scientific theory, is an unconscious derivative from medieval theology".[31] It is also true, however, that medieval theology, based as it was upon Aristotelian thought would certainly have postponed the development of modern science. Had the ingredients of medieval theology not been overcome they may well have prevented the emergence of modern science.

Over against Whitehead it must be emphasized that, while Aristotelian thought was essential for the orderly classification of nature, it also worked against the investigation of nature for its own sake. Aristotle's deductive method, whereby the nature of reality was logically deducible from presupposed general principles, placed thought structures in a kind of straitjacket that had to be torn away before experimental science could develop.

Grosseteste, Scientist at Oxford
While it is true, as A. E. Taylor (1869-1945) has pointed out, that in England "Thomism never really became at home, the Augustinian tradition, combined with a spirit of devotion to mathematical and experimental science, was ardently cultivated all through the centuries by the Franciscans of Oxford, Robert Grosseteste, Roger Bacon, and their friends".[32] Nevertheless the influence of Arab Aristotelianism is not to be denied even at Oxford. Grosseteste not only knew the works of Aristotle as translated into Latin from Arabic, but he wrote commentaries upon those works and translated Aristotle's *Ethics*. This would give a different twist to matters than what A. C. Crombie implies with his statement that Grosseteste "was able to give Augustinian-Platonism a twist which turned the inquiry for God in things into the first systematic experimental investigations of things [as such]".[33]

Grosseteste himself refers to Arab commentators of Aristotle, especially Averroës,[34] and uses both Augustine and Aristotle in working out his understandings of the world. Pause, therefore, ought to be given to any argument that would put too much emphasis on Grosseteste as a follower of Platonism or Augustinianism alone. The fact that his interest and writings on Aristotle when he was at Oxford pre-date those of Albert the Great in Paris may indicate that it was he, rather than Albert the Great, who was first responsible for introducing Aristotelian thought into the realms of western academia.[35]

Grosseteste began his scientific investigation by formulating a neo-Platonic metaphysics of light. Plato, by way of Socrates (c. 469-399 B.C.), had long since argued that the supreme God of the heaven was the sun. It was the source of light, fire, generation, nourishment and growth. It was also the light by which the sun itself and all that it illuminates was seen.[36] For Augustine, following Plato and Plotinus, God was infinite and incorporeal light. At the same time God was the source of incorporeal and corporeal created light, and created light was the noblest among created bodies.[37]

Grosseteste, however, moves from metaphysics to physics. For him light was a "corporeal form", the substantial basis of spatial dimensions as well as the first principle of motion. Thus, the laws of light were nothing less than the foundation of the scientific explanation of the world.[38] Following the kind of neo-Platonic thinking evident in both Plotinus and Proclus and as emphasized by the place and energy of the sun in Hermeticism, light for Grosseteste was fundamental to both motion and change. "When light generates itself in one direction drawing matter with it, it produces local motion (*motus localis*); and when the light within matter is sent out and what is outside is sent in, it produces qualitative change (*alteratio*)".[39]

It is with regard to light that Grosseteste, like Philoponos, emphasized the doctrine of *creatio ex nihilo*. *In doing so he sided with Augustine against Aristotle.* Augustine insisted that God created all things — the heavens and the earth, the spiritual and the corporeal — out of nothing.[40] Likewise, for Grosseteste God had created unformed matter (*materia prima*) out of nothing at the beginning of time; and he maintained that both light and matter first appeared at creation. Nonetheless, in a Platonic and neo-Platonic manner, light (*lux*) played an instrumental role in creation. In diffusing itself, it produced both the dimensions of space and all subsequent things.[41] In contrast to the neo-Platonists, however, for whom the sun and light were divine substances, Grosseteste, again like Augustine and Philoponos, insisted that they were created substances that were subject to study. Yet he retained the idea that light was primary. Thus, Crombie is quite right in stressing the fact that "Grosseteste believed that the study of optics was the key to the understanding of the physical world".[42]

The next and necessary step was the application of Pythagorean mathematics to investigate light and mark the incidence of its angles. He thus moved beyond Plato's concept of mathematics as a means of contemplation and, like Euclid and Ptolemy, utilized mathematical geometry to study and explain the behaviour of light. Since the late Pythagoreans, exemplified by Ptolemy, had long since measured the phenomena of the heavens, Grosseteste began to measure earthly phenomena. He did this by combining mathematics, as advocated by Plato and used by the later Pythagoreans, with the inductive procedures of Aristotle's *Posterior Analytics*.[43]

Grosseteste set out a "scientific method" that was to set a precedent for experimental science. The method was continued by his student, Roger Bacon, and influenced both Duns Scotus and William of Ockham. It is especially worthy of note that, while Grosseteste had studied both Plato and Aristotle, he modified and used selectively their teachings, especially as voiced by Augustine, in the light of the Christian faith. His insistence on the reality of the material world, use of induction rather than simply deduction, and concentration on measurable particulars rather than upon ubiquitous universals stands at the very foundation of the modern scientific method. Eventually, the combination of inductive procedures and mathematics was to lead to the break-down of the Aristotelian deductive rationalism of the late Middle Ages. It prepared the way for the kind of quantified experimental science that was to blossom in the seventeenth century.

As important as the doctrine of *creatio ex nihilo* was for Grosseteste, numbers remained for him, as indeed they had been for the Pythagoreans, eternal. They were differentiated from matter as the soul was differentiated from the body. Mathematics was the eternal, abiding element in all transitory processes. He makes his position clear in his commentary on Aristotle's *Physics*. He admits that he follows Aristotle in differentiating between physics and mathematics; and likewise distinguishes between physical bodies in motion and the magnitudes abstracted from them. Magnitudes as such, cannot be considered properties of physical bodies. Rather, they represent the relationship of moving bodies to one another.[44]

For the early Pythagoreans and Plato, of course, mathematics was the essence of reality. Numbers combined as ratios and formed the geometrical patterns of the circles which described the eternal heavens and the essence of the earthly elements. The cube was the essential reality of *earth*, the icosahedron of *water*, the octahedron of *air*, and the tetrahedron of *fire*. These forms were the reality of the world in mathematical form.[45] Grosseteste agreed with the later Pythagoreans such as Aristarchus of Samos (c.320 - c.250 B.C.), Eratosthenes (c.275 - c.195 B.C.), Hipparchus (fl. 145-127 B.C.), and Ptolemy (A.D. c.96 - c.168) that mathematics was eternal and unchanging. It provided a standard to which natural phenomena could be referred in order to stipulate similarities and differences in terms of numbers, angles and ratios.[46]

Although Plato had allowed mathematics to span both heaven and earth in a theoretical way, Grosseteste's stress on *creatio ex nihilo* allowed him to see the whole of nature, the heavens and the earth, as of a piece. The unity and harmony of nature, advocated by Plato and the neo-Platonists as being attributable to the *divine essence* of the world, was for Grosseteste a characteristic of the whole of natural reality. Nature's oneness or unity gave it the kind of uniformity which allowed it to be subject to mathematical quantification as well as to the law of parsimony.

The idea of material uniformity was known to a limited extent as early as the pre-Socratic Ionians who had identified the heavenly bodies as being of earth-like substances. Later, however, the Pythagoreans and, following them, Plato and Aristotle, re-divinized the heavenly spheres and made the unity of nature dependent on the divine world-soul. The law of parsimony (*lex parsimoniae*) or the principle of economy, which was to become known in the West as "Ockham's razor", goes back to the early Pythagoreans. From them it was inherited by both Plato and Aristotle and given expression in their insistence that the circular motion of the heavens was the simplest and most perfect of all motions.[47]

In the Christian West, interestingly enough, the uniformity of created reality, or as we have learned to call it, the "isotropy of nature", had begun to be stressed from a theological perspective as early as Clement of Rome (fl. A.D. 96). Clement placed all of nature in a contingent relationship to God and saw it as being a uniform, single, interrelated structure. Hence, in his *Epistle to the Corinthians*, Clement insisted that the world of the antipodes was directed "by the same ordinances of the Master" as the inhabited world.[48]

Grosseteste continued this emphasis on nature. He stated the law of parsimony as, "Nature operates in the shortest way possible but the straight line is the shortest of all".[49] He may have advanced the scientific cause to an even greater degree when he advocated the relationship between the nature of things and their effects as set out in his *De Generatione Stellarum*.

Things of the same nature are productive of the same operations according to their nature. Therefore, if the same operations are not produced by their natures, they are not of the same nature.[50]

Here *in nuce* is expressed the entire basis in modern physics for identifying objects in terms of their interactions. It was with regard to the *nature of things* that Grosseteste, like Philoponos before him, challenged Aristotle and the whole of Greek cosmology from the Pythagoreans onward. Whereas the Greeks with the exception of the Ionian *physikoi* held that the heavens were simple, Grosseteste argued that both observation and logic indicated that the heavenly bodies, like those on earth, were complex bodies. They were mutable rather than being either eternal or incorruptible.

Grosseteste's argument for the complex nature of the heavens is two-fold. He first insists that the visible heavenly bodies are of a different nature than the putative unseen spheres to which, according to both Aristotelian and Ptolemaic astronomy, the bodies were attached for their motion. Bodies that appear different have different natures.

Likewise, bodies that effect sight are therefore of a different nature than bodies that do not.

Grosseteste followed Philoponos in identifying the heavenly bodies as being complex rather than simple. According to Aristotle's *Physics*, colored bodies were complex. Since stars obviously had color, it followed that the nature of the stars themselves had to be complex. In this way Grosseteste used Aristotle against himself. Again Grosseteste pursues Aristotle's argument that bodies move according to their natures for his own purposes. Whereas Aristotle held that the simple nature of the heavenly bodies corresponded to their simple and regular circular motion around a single center, Grosseteste argued that since it was now evident that the movements of the heavenly bodies were complex and they did not move concentrically, the bodies themselves are complex.[51] He extended the argument by showing that since, following Aristotle's logic, the heavenly bodies were necessarily complex and complex bodies were, again by Aristotle's definition, temporal and changeable, the heavenly bodies were neither eternal nor immutable. This, of course, paralleled his conclusion that both heaven and earth were of created nature.

Grosseteste's theory of the uniformity of nature was the groundwork for his scientific method and his application of geometry both to astronomy and optics. This theory subjected the entirety of nature to the same laws including the law of parsimony which allowed for the processes of selection through exclusion and contradiction. Therewith he was able to calculate the variations in the observed phenomena of the heavenly bodies as well as to measure and differentiate between angles of both incident and reflected light.

In his writings on astronomy there is no indication that Grosseteste had knowledge of Aristarchus' heliocentric world or of the cycles, epicycles, eccentrics and equants of Ptolemy. His diagram of the world followed the usual Aristotelian-inspired geocentric homocentric design with the earth in the center surrounded by the spheres of the other three elements—water, air and fire. Around these he placed in consecutive order the spheres of the moon, Mercury, Venus, the sun, Mars, Jupiter, Saturn, and the firmament of the stars (*coelum stellarum*). The whole, somewhat anomalously, was surrounded by a ninth sphere (*sphaera nona*), the *prime mover*.[52] I say "somewhat anomalously" because in both his *De Sphaera*, his primary text on cosmology and in his metaphysical arguments Grosseteste follows Plato rather than Aristotle with regard to motion. Motion, for them is not dependent upon a "prime mover", but upon a "world soul".[53]

A more blatant inconsistency in Grosseteste's writing is that he both agrees with Aristotle that the heavens are composed of a "fifth essence" (*quinta essentia*) and, as we have seen, contradicts Aristotle when he argues that the nature of the heavens was complex rather than simple. In addition he correlates the complexity of the movements of the heavenly bodies with the complexity of their natures.[54] He extends his disagreement with Aristotle by insisting that the heavens shared the diversity of creation and were also subject to permutation and change.[55] They were of created nature and of finite motion. All of this was, of course, rank heresy to the Aristotelians who devotedly believed that the heavens were immutable and eternal.[56] Nevertheless, Grosseteste continued to follow Aristotle in attributing the spherical shape of the world to circular motion.[57]

It is in his discussion of motion, *De motu supercaelestium* (*The Movement of the Highest Heavens*) that Grosseteste calls Aristotle's concept of a "prime mover" into question. It was obvious to Grosseteste that Aristotle's concept of a prime mover who imparted *simple* motion to the heavenly bodies, could not explain the complexity of movement to which the heavenly bodies were observed to be subject.[58] In addition, there was a double difficulty in Aristotle's insistence that both time and the prime mover were eternal. Grosseteste's primary objection was that the idea of created things being eternal was contrary to the Christian doctrine of *creatio ex nihilo*. Secondly, and this is especially telling, Grosseteste saw that Aristotle's argument involved a logical fallacy.[59]

While Aristotle had insisted that all movement presupposes antecedent movement and an antecedent mover, he also argued that the movement in question was without beginning and end.[60] This meant that, at one and the same time, Aristotle subscribed to the existence of a prime mover that imparts first motion, and to the eternality of motion, to which there could be no first movement. Hence, Aristotle was obviously caught in a contradiction of his own making.[61] It was thus more reasonable for Grosseteste to move back to Plato and to attribute the complex motions of the heavens to a divine soul.[62] Thus, although Grosseteste both adopts certain aspects of Plato and Aristotle's conception of the heavens, he disagrees with both their understanding of the composition of the heavens and their understanding of motion. He was convinced that the heavens, like the earth, were finite, generated, temporal, and subject to change and corruption.[63]

Grosseteste's utilization of mathematics to explain observed phenomena is especially evident in his *Optics*. He shows that the incident rays and those reflected in a mirror form equal radiant angles. Further, these correlated with the Euclidean geometrical formula stating that if in comparing any two triangles the angles of the first triangle are equal to an angles of the second triangle and if the sides forming the equal angles are proportional, then the other two angles of the triangles will be equal as well. In addition, Grosseteste's observation enabled him to formulate the law of reversibility. "Since the operation of nature is finite and regular, the path of regeneration must (*necesse est*) be similar to the path of generation, and so it is regenerated at an angle equal to the angle of incidence."[64] The law on which cause, effect, and predictability was to be based became the *sine qua non* of Newtonian physics.

Grosseteste demonstrated that mathematics was not only a medium for describing the operations of nature according to a system of numbers, ratios and angular measurement, for it could be used to define operations performed by experiment. Following "the principle of natural philosophy that 'every operation of nature takes place in the most perfect, orderly, briefest and best way that is possible'" (the law of parsimony),[65] Grosseteste explained the behaviour of light on the basis of mathematics. He then applied laws pertaining to the refraction of light to other optical phenomena such as the operation of the spherical lens of a burning glass,[66] and the different angles formed by the rays of light as they passed from a rare to a dense medium.

Basing his theory on the supposition that there was a correlation between the density of the medium and the angles of the incidence of refraction, he attempted to reduce the colours of the rainbow to quantities. He quantified transparency in relationship to purity and impurity or cleanliness and dirtiness and measured light according to clarity and darkness, abundance or rarity. The quantifiers had their causes primarily in the abundance or paucity (*paucitas*) of the rays of the sun and were thus supposed to account for the different colors that resulted.[67]

Grosseteste's explanation is in the end erroneous. However, his correlations between the density of the medium through which light passed and the increased angle at which it was refracted along with his observation that there is a correlation between the refraction of light and its color were steps in the right direction. Proper correlations were made by Newton some 450 years later when he refracted sunlight through prisms. Grosseteste's detailed documentations were significant steps in the correlation of phenomena and numbers that form the very basis on which experimental science is built.[68]

Impressive as were Grosseteste's early attempts at instigating and implementing a scientific method, it is perhaps even more important that he understood Aristotle well enough to be selective in the use of his ideas. While he adopted the inductive logic of the *Posterior Analytics* he rejected the emphasis in Aristotle that insisted on the idea of a single, necessary truth from which all truth could be deduced and developed.

Instead Grosseteste instituted a concept of what we may call "levels of truth". The inspiration for this may well have come from Anselm of Canterbury (1033-1109).[69] To be sure,

God is truth in an absolute sense but, like Einstein who insisted that reality exists independent from our observations, Grosseteste insisted that created things have a truth of their own, a truth independent of the truth of proposition (*veritas enuntiationis*) by which they are signified. The truth of things on their own is also quite different from the highest truth (*summa veritas*), the truth of God from which the truth of created things is derived. Following Augustine, who insisted that it is only as the mind is enlightened by the highest truth that it is endowed with intellectual vision.[70]

Grosseteste was certain that the truth of things may be known only in the light of the highest truth. Nevertheless the truth of nature is not sacramental, a reflection of the divine, as it was for Augustine. Rather, the truth of nature is *sui generis*. This is clear from Grosseteste's explanation of change in created nature wherein he insisted that a fundamental difference must exist between the truth concerning the essence of created things (*veritas essentiae rerum*) and the truth of God. Hence, in contrast to Platonism, Aristotelianism and neo-Platonism, Grosseteste desacralizes nature. The truth of God is one thing, that of nature is another. It was precisely this recognition of truth as *essential to things* in their own right that made knowledge by experience and knowing by experimentation possible.

However, Grosseteste does not leave God out of his explanation. Rather, as Ludwig Baur (1871-1943) has demonstrated, Grosseteste was able to conceive the truth of things as different from but contingent upon eternal truth. He claimed, for instance, that as water is held in a square form only insofar as it is surrounded by a square vessel, so the created world (without being *held*, as it were) must be regarded in itself as tending to slip into non-being (*labile in nonesse*), simply because the world lacks self-sufficient reality or substance in itself.

Although in the first instance Grosseteste's understanding in this regard relates to the presence of the soul, which as in Plato gives shape to reality, as well as to the "form-substance" doctrine of Aristotle, his strict differentiation of

God from the world shows that the thrust of his concepts is away from the identity of God with the world and toward a concept of contingence of created reality upon God. In their utter differentiation from God, created things continue to exist only insofar as they are unceasingly upheld or supported by God's eternal Word (*supportari ab aeterno verbo*). This Christian understanding of created being, in which the grace of God is dynamically related to the world's reality yet differentiated from it, allowed Grosseteste to use Aristotelian and Platonic thought and, at the same time, to break with it. This break came at the point where its monolithic structure necessitated that the world be understood as essentially divine and interpenetrated with first and final causes.

The same insight which allowed Grosseteste to differentiate between levels of truth permitted him to assign different values to mathematical and historical truth. Only the former was certain for, according to Grosseteste, the latter was subject to interpretation. Accordingly, contingent and developing entities necessitate a different perception than do ideas that are purely conceptual. Somewhat like Descartes later, we are more certain of the latter than the former.

This indicates that Grosseteste's ideas about knowing the future do not belong to the Aristotelian rationalism that believed itself capable of divining the end from the beginning. A judgement about the future only becomes true for Grosseteste when the *supposition* made in the present about the future is followed by a *future actuality* which agrees with the supposition in question. The *veritas de futuro* (truth about the future), therefore, demands a *secundum quid* (something following). Further, since the *secundum quid* is itself contingent with regard to future circumstance, it *ipso facto* can never have the status of actuality in the present. Hence, human perception is limited by time. It will remain limited until "our spiritual eye is cleansed of its temporal make-up (*temporis compositione*) and raises itself [in a somewhat neo-Pythagorean and neo-Platonic way] to contemplate pure eternity".[71] Accordingly, our understanding of the world is inevitably limited by the constraints of time itself. Our knowledge of reality cannot be a matter of deduction from eternal principles but must concern itself with things in their contingent relationships. This may very well work out in a way that we do not expect or foresee.

Grosseteste's work demonstrates that in the early twelfth century, Aristotelian thought, though helpful in the development of science, had not become powerful enough to dictate either scientific or theological thought. A generation later, however, it was on its way to becoming the *norma normanda* for both the theology and science of the medieval church. Yet in a real sense the latter phase of dogmatic Aristotelianism can be seen as an infinitesimal reaction that both held the development of science in check and attempted to reverse it. It was to use of Aristotle not wisely but too well. With Grosseteste, however, Aristotle's ideas are used selectively. They are appreciated, utilized and/or compromised, contradicted and ignored as seemed propitious to the method of observation, theory, and experimentation that eventually developed and grew into modern science.[72]

Although in the fifth century Philoponos used Aristotle selectively, contradicting him when it seemed warranted, he did not revolutionize the scientific endeavor. As Crombie has noted, it was with Grosseteste that things began to change. At least with regard to the modern (post-medieval) development of science, Crombie is right that, with Grosseteste, Oxford became the first center for the methodological revolution that led to modern science.[73] Grosseteste appears to have been the "first medieval writer to recognize and deal with the two fundamental methodological problems of induction and experimental verification and falsification which arose when the Greek conception of geometrical demonstration was applied to the world of experience".[74] As we shall see at a later stage, it was in fact the conscious insertion of the logic of induction into the intellectual milieu which had become dominated by Aristotelian deductive logic that both undermined the structures built on Aristotelian rationalism and, in doing so, opened the way for both the Protestant Reformation in the sixteenth century and the Scientific Revolution in the seventeenth century.

Roger Bacon's Anti-Aristotelian Science
A sometimes neglected figure in the history of the development of science is the Franciscan Roger Bacon (c.1214 - c.1294). Bacon had learned both the rudiments of mathematics and the scientific method from Grosseteste at Oxford. At that time, in the words of Robert Adamson (1852-1902), Oxford was both the seat of opposition to papal supremacy and the place where "the whole current of thinking had a tendency away from traditional learning towards mathematical and experimental science".[75] From Oxford Bacon traveled to the University of Paris just at the time Aristotelian thought was well on its way to captivating European thought especially through the teaching of Albert the Great and his student, Thomas Aquinas.

As Crombie points out, Bacon, like Grosseteste, began his theory of science with Aristotle's *Posterior Analytics*. Bacon also followed Grosseteste in using the logic of induction from experiment to arrive at universals and in using mathematics to show the quantitative relationships between phenomena.[76] Thus, Bacon, like Grosseteste, was heavily dependent on Aristotle for the rudiments of his science and especially for his cosmology, his physics and his astrology.[77] He has every reason to call Aristotle "the wisest of Philosophers".[78] Yet he was far from being a convinced Aristotelian. He set his two Oxford teachers, Grosseteste and Adam Marsh (d. c.1258) on a par with both Aristotle and the Arab Aristotelian, Avicenna, and claimed, perhaps with some exaggeration, that they were perfect in both science and philosophy.[79]

Like Grosseteste, Bacon both questioned Aristotelian thought at crucial points, and pointedly rejected his deductive system and the authoritative structures it engendered. That rejection eventually reached a point of passion. In the end he openly defied Aristotelian teaching. He blatantly faulted those who were so foolish as to try to gain scientific knowledge by way of deduction from accepted authority rather than by way of the investigation of things through "experience". Most likely, it was Bacon's anti-Aristotelian stance, perhaps along with his "experiments" in the "black art" of *al chimie* (the art of the land of *Khem*, which is Arabic for Egypt), that were the grounds on which he eventually became a *persona non grata* with his Franciscan superiors. According to what would seem to be a fair reading of the evidence, he was imprisoned for fourteen years.

The various reports of Bacon's imprisonment seem to indicate that he was most likely "sentenced" by Jerome de Ascoli (1227-1292), who had become General of the Franciscan Order in 1274. Four years after his promotion to the head of the Order, Ascoli "incarcerated" Bacon for certain "novelties". After the death of Ascoli, who had become Pope Nicholas IV, Raymond de Gaufredi (d. 1310), Ascoli's successor as General of the Franciscan Order, reconsidered the sentences of 1278 and set Bacon free along with others who were imprisoned with him.

Whether or not the "novelties" for which Bacon was imprisoned included his practice of alchemy, as a note in one manuscript indicates, and whether or not Thomas Bungay (fl. 13th cent.) with whom he may well have shared "alchemistic secrets", helped affect his release, as Adamson testifies, is not certain. Yet it is certain that Roger Bacon, along with a host of late medieval, Renaissance and even post-Renaissance scholars who were interested in the interrelationships of spirit and matter, was an ardent practitioner of the "black art". Like all medieval alchemists, he hoped not only to produce gold of any purity desired but also medicines which would prolong life.[80]

Today, for the most part, we have learned to look upon alchemy with a measure of scorn. We forget that its practice, stimulated in the West by Robert of Chester's (fl. 12th cent.) translation of *The Composition of Alchemy* from Arabic into Latin in 1144, initiated the establishment of the first laboratories of late medieval Europe. Built on the rationale of emulating "natural processes" at an accelerated pace, their chemical experiments attempted to transform sulphur and mercury into gold and to distill the "fifth essence", the "elixir of life" from the natural elements. The fires under their pots was matched only by that of their ardor as they reached beyond their grasp to manipulate nature to give up her secrets.[81]

The alchemistic experiments that span the centuries from ancient Egypt to the seventeenth century and even into the eighteenth, enticed not only Roger Bacon but more famous figures such as Francis Bacon (1561-1626) and Isaac Newton (1642-1727).[82] Roger Bacon's interest, like that of other alchemists, was without doubt spurred on by the dream of discovering the formula for "the philosopher's stone". It was most likely out of the experience of his alchemical investigations that Bacon's attitude towards science, including its application arose.[83] Eventually alchemy developed into chemistry. We have all but forgotten that its etymological roots are Egyptian and its original motivation was grounded in the perfection of the "black art".

At the time Bacon migrated from Oxford to Paris, the university was dominated by the thought of Aristotle that reached the western mind through the writings of the Arab scholars, Avicenna, Alhazen [Ibn al-Haitham] (965 - c.1040), Averroës and Alghazali. Here its reception was much more uncritical than was the case at Oxford.[84] In Paris Aristotelianism was both represented and promoted, although not altogether uncritically, by Albert the Great and by his renowned pupil Thomas Aquinas. Through Thomas, who more than any other, was to be responsible for incorporating Aristotle into medieval thought, Aristotelianism became the basis for a vigorous, comprehensive, and within the constraints of its own logic, thoroughly rational system of knowledge. As defined particularly by the *Prior Analytics* and as complemented by the *Metaphysics* and *Physics*, Aristotelian deductive rationalism became through Thomas the basis of the dominant medieval theological system. In due time Thomas' appropriation of Aristotle became the touchstone of both ecclesiastical and scientific orthodoxy. The system that moves from "the truth" of known principles to "the truth" of particulars with inexorable logic, is understandably persuasive. In its own way it embodies simplicity

The Christian Critique of Aristotle

to such an extent that even today a great deal of Roman Catholic and Protestant theology continues to be pushed along its straight-forward path.

Early in this academic career Bacon was influenced by Aristotle. During his sojourn in Paris, he nonetheless shows contempt for the kind of thinking which depended on authority and deduction as the way to truth. His disdain for the all-encompassing deductive system that bore the name of "Aristotelianism" in Paris, carried over to his scorn for Thomas Aquinas whom he called a *"vir erroneus et famosus"* (a mistaken but celebrated man).[85] By contrast, Bacon referred to Petrus Peregrinus de Maricourt (fl. 2nd half of 13th cent.), "a master of experiment", who inspired him to move in the direction of *scientific procedures.* In words that are doubtless too complimentary, but that also belie Bacon's, and by implication de Maricourt's familiarization with Hermetic alchemy, Bacon said of de Maricourt, "By experiment he knows the things of nature and medicine and alchemy and all that is in the heavens and below".[86] It is likely more than coincidental that the Hermetic axiom used in support of alchemy reads: "What is below is like that which is above and what is above is like that which is below, to accomplish the miracle of the one thing."[87]

The scholastic method was practiced by those who, according to Bacon, "shut their eyes to nature", and deduced things by argument and logic whether from Aristotle or the Bible. For Bacon, however, the only assurance of truth was that it be correlated with "experience". Experience, however, had to be quantified. As we have seen Bacon not only praised Grosseteste for his mathematical knowledge,[88] but he was convinced that mathematics was indispensable to "science" as he knew it.

> He who is ignorant of this [the science of mathematics] can not know the other sciences nor the affairs of this world ... the knowledge of this science prepares the mind and elevates it to a certain knowledge of all things, so that if one learns the roots of knowledge placed about it and rightly applies them to the knowledge of the other sciences and matters, he will then be able to know all that follows without error or doubt, easily and effectually.[89]

He then notes the three essential parts of philosophy which Aristotle mentioned in the sixth book of his *Metaphysics*: the mathematical, the natural and the divine. "The mathematical", Bacon says in somewhat of an understatement, "is of no small importance in grasping the knowledge of the other two parts".[90] While he points out that grammar and logic were quite indispensable to knowledge, Bacon was certain that only in the realm of mathematics was there "true and forceful demonstration".[91] Mathematics was equally pertinent to things celestial and things terrestrial.[92] It applied to practical astrology [astronomy] which "enables us to know every hour the positions of the planets and stars, and their aspects and actions, and all the changes that take place in the heavenly bodies".[93]

For Bacon the correlation between mathematics and nature was built on his understanding that Aristotle's [actually the Pythagorean] principle that nature "works in the shortest way possible" was translated from Euclid's "straight line." The result was that the forces of nature could be calculated according to the lines and angles of geometry.[94] Bacon portrayed the use of mathematics in his optical demonstrations. He computed the angles of refraction from the sunlight through a goblet of water,[95] the refraction of light from a dense medium,[96] and the behaviour of light with regard to spherical surfaces.[97] The sphere, for Bacon as for Aristotle was nature's favorite figure. He was certain that all geometrical figures could be inscribed within it. "It is this figure that nature especially selects in every multiplication and action."[98]

The emphasis at Oxford on the importance of mathematics can be judged by Bacon's mastery of Euclid's *Elements*. With it he explained the relationships of angles and sides of triangles to one another. He demonstrated the conical shape formed by the sun's light rays falling upon the earth and the differentiation which occurs in the incident angles of light that depend upon the sun's position. He pointed out the pattern of the incidence and reflection of the sun's light on and from the surface of the moon on its way to the earth. He calculated the mathematical values of different astronomical phenomena as observed from different parts of the earth and at different times of the year.[99]

Preceding Newton by some 400 years, and contrary to Galileo's mistaken belief that the tides were caused by the combined rotation and revolution of the earth, Bacon showed that the ebb and flow of the sea was due to the position of the moon. Yet Bacon did not discern that the moon's effect upon the tides was due to its gravitational attraction. Rather, he explained the ebb and flow in relationship to the different angles of incidence of the moon's rays. The explanation, which is more important for its ingenuous use of geometry than for its factuality, is commensurate with Bacon's astrological ideas. These included the supposition that right angles were of greater force than oblique angles. When the moon was directly overhead, it drew a sufficient amount of vapor from the water to hold the sea in check. However, when the moon approached the meridian, the rays falling at an oblique angle were weaker. The weaker rays were able only to draw the water from the depths of the sea and this allowed the sea to overflow on to the land.[100]

In a combination of mathematics and myth, truth and fantasy, Bacon attempted to correlate the angles of light rays with their power. He warned that, for the sake of the health of the body, one should avoid the "principal forces of harmful things, namely, those refracted and reflected in straight lines". Especially "when a man is exposed to harmful celestial impressions, like the sun in summer and the moon at night, which exhaust our bodies" one "should seek out the accidental forces evading their fall at a right angle" and be aware at least of the "shorter rays of the pyramid".[101] Fanciful as were many of Bacon's deductions, they indicate the seriousness of his attempts to apply mathematics to the whole of nature, astronomy, geography, geophysical relationships and human health.

He came much closer to what we recognize as scientific procedure when, following his master Grosseteste, he turned to optics. As noted earlier, like the Greeks before them, Bacon as well as Nicholas of Cusa and Copernicus, considered the sphere to be "nature's favorite". It is hardly surprising, therefore, to find that he opens his discussion of optics with an explanation of the structure of the eye drawn as a series of non-concentric spheres. He adopted the anatomical arrangement from Alhazen and Avicenna. He explained that when light enters the eye, it is refracted through the various parts and humours until it is focused on the "whole optic nerve". The optic nerve is "the point of ultimate perception". It not only transmits the physical objects to the brain but, following Alhazen, is itself the anterior part of the brain.[102]

Here again, as with his attempts to use mathematics to explain the tides and the other forces of nature, Bacon was not completely successful. He endeavored to explain both the structure and function of anatomical organs by "mapping" them in mathematical terms. Admittedly his diagram was schematic. It was a drawing "on a surface". "The full demonstration would require a body fashioned like the eye in all the particulars aforesaid." In other words, mathematics was used to demonstrate the object under investigation but it was the *object* on which the focus of investigation must concentrate and to which mathematics must conform. In the case of the eye, the experiment could proceed, Bacon suggested, by illustrating the process with "the eye of a cow, pig, and other animals".[103]

With this kind of allusion to experimentation one begins to see the basis for Bacon's repudiation of Aristotelian procedures of deduction. As with Grosseteste so with Bacon the impetus may well have come from Plato who urged that inquiry into nature be made by using hypotheses.[104] Granted that for Plato *true knowledge* could be given only by revelation, it was manifest in geometry and arithmetic as well. However, it was when one adopted starting points concerned only with the shadows of reality that the scientific procedure could begin. In contradistinction to Aristotle, therefore, Plato insisted that the hypotheses were not "first principles". Rather, they were

> steps and points of departure into a region which is above hypotheses [utilized] in order that she [the soul] may soar beyond them to the

first principle of the whole; and clinging to this and then to that which depends on this, by successive steps she descends again without the aid of any sensible object, beginning and ending in ideas.[105]

Plato is quite clear that the ideas or concepts gained by the use of hypotheses are of utmost importance. As Taylor points out they are supposed to "exhibit direct self-evidence"[106] "without the aid of any sensible object",[107] and "making no use of images.[108] Bacon, like Grosseteste, Duns Scotus and William of Ockham, all of whom had been nudged toward considering the importance of nature by the works of Aristotle, used hypotheses to project what nature itself might possibly consist of. The hypotheses, then, could be tested for validity.

It is in this light that Crombie says of Bacon that his procedure "represents the first major advance made in the experimental method after Grosseteste".[109] Bacon insisted that it was only by experimental science that one could differentiate between truth and illusion. Experimental science alone,

> knows how to test perfectly what can be done by nature, what by the effort of art, what by trickery, and what the incantations, conjurations, vocations, depreciations, sacrifices, that belong to magic, mean and dream of, and what is in them, so that all falsity may be removed and the truth alone of art and nature may be retained.[110]

This method is nicely illustrated by his attempt to explain the shapes and colors of the rainbow. Bacon knew that both Aristotle and Avicenna had discussed the rainbow. He was certain, however, that

> neither Aristotle nor Avicenna in their Natural Histories has given us a knowledge of phenomena of this kind, nor has Seneca, who composed a special book on them. But Experimental Science attests them.[111]

Bacon's method combines observation, induction toward hypothesis and deduction from them. In the case of the rainbow his experimental procedure is outlined as follows: First, the experimenter should examine visible objects with the color and the figure of the rainbow and attempt to determine which characteristics tend to produce the colors. Second, notice should be taken of the same phenomena that occur in a variety of other circumstances: water falling from the wheels of a mill, drops of dew on the grass on a summer's morning, rays of sun falling through the rain or the sun's rays itself as seen from a peculiar angle. Third, one inquires, "What do these phenomena have in common with the rainbow?" The answer will indicate the common cause of the diverse phenomena.

When he investigated the shape of the rainbow, Bacon correlated the observation of different instances, and surmised that there was a relationship between the height of the bow and the altitude of the sun. Utilizing his knowledge of optics wherein incident rays of the sun formed a conical shape as they came to earth, he deduced that the bow-shape was given by the fact that it represents the top half of the circumference of the base of a cone, the apex of which was the sun and the axis of which passed through the observer's eye to the center of the bow.[112] The position and size of the bow was dependent upon the altitude of the sun in relationship to the position of the observer.

Bacon nicely illustrates this scientific procedure in his attempt to discern the cause of the rainbow. His repertoire included three hypotheses: incident rays, reflection or refraction, and whether or not, as he says, "there are real colors in the cloud itself".[113] He drew his conclusions by allowing the different hypotheses to be confirmed or rejected on the basis of observation. Experience had shown that the rainbow moved at an equal speed to the individual moving toward or away from it. Thus, Bacon concluded that the rainbow was an optical phenomena that appeared in relation to the individual observer. There were, in fact, as many rainbows as there were observers.[114] In that the rainbow moved with the observer, it was obvious, Bacon pointed out, that the rainbow could be caused only by the reflected rays of the sun. The rays were reflected by the raindrops in the depth of the cloud. Each drop had the "nature of a mirror".[115] Rather than being integral to the cloud itself, the colors were merely a matter of vision.[116]

Therefore, real color is nowhere present in the moist cloud, nor is the appearance of color present except in the drops from which reflection comes to the vision.[117]

Bacon then correlated his conclusions with the phenomenon of the circles that appeared under certain conditions around a candle. These were thought either to be formed by vapors set free from food or drink, from moist air, or produced at a distance from a candle in the vapors of the moist eye. He ended his argument by reiterating his stance, "Any scientist will readily admit that experiment, not reasoning, determines the conclusions in regard to these matters".[118]

This was *science*, rudimentary in its method, perhaps, but *science* nevertheless. It applied first of all to knowledge of nature. It is important to note, however, that, as far as Bacon was concerned, the ramifications of the scientific method went far beyond knowledge of nature. It was, in fact, the basis of knowledge in general. Such a method was even necessary for understanding the Bible. Knowledge of the Bible required first of all the ability to use the original languages. In addition, however, it was only by the use of a proper understanding of the "sciences" that the innumerable references to natural phenomena and to handicraft with which the Bible abounds could be made intelligible.

H. S. Redgrove's (1887-1943) statement that because Bacon "lived in an age dangerous for men of speculative mind", interest in the sciences was necessary "primarily for the full and complete understanding of theology, queen of them all",[119] is misleading. It is more likely that Bacon was serious about the general application of the scientific method. For instance, in that Bacon remained beholden to Aristotle's conception that nature was a reflection of divine forms, he remained convinced that the knowledge of God could be read from the world. Therefore, like the early Pythagoreans and like Kepler and Newton, Bacon was convinced that to be enabled by way of science to know the world was to be enabled to know God. Whether this *deus sive natura* conceptuality is of the Aristotelian or of the Platonic neo-Pythagorean genre would seem to be an open question. Nevertheless, his conviction in this regard underlines Bacon's insistence that natural science supported theology.

The matter is compounded in Bacon by his partial adoption of the "science of Hermeticism". He believed it supported theology along with "legitimate science". Hermeticism was a strange combination of gnostic, cabbalist, occult, neo-Pythagorean and neo-Platonic mystical ideas masquerading as ancient religious, philosophical and scientific truth. While never officially sanctioned, it influenced the medieval mind far beyond the value of its actual doctrines. Like the original Pythagorean cult, it was secretive in nature. It can be gauged as much, if not more, by its secondary effects as by the outright admissions of those who were influenced by it. Nevertheless, Hermetic ideas seem not only to have influenced the Renaissance outlook in general, but to a greater or lesser degree served to form the minds of such well-known thinkers as Nicholas of Cusa, Nicholas Copernicus, Giordano Bruno, Johannes Kepler and Francis Bacon. It is with Roger Bacon that we seem to have our earliest evidence of the contribution of Hermetic ideas in the development of science.

As Lynn Thorndike (1882-1965) points out, Bacon was well acquainted with Trismegistus Hermes, the thrice-great, who according to Hermetic lore was the "father of philosophers".[120] Trismegistus was held to be somewhat younger than Moses but older than Plato. Thorndike's conclusions that Bacon was caught up in the Hermetic traditions seems to be supportable. This may indicate that Bacon's disdain for Albert the Great was not entirely due to Albert's adoption of Aristotelian rationalism. Bacon's antipathy for Albert may also be due to the fact that in his *Speculum Astronomiae,* Albert had condemned Hermeticism as "the worst kind of idolatry".[121]

As we noted above, there are good, very good reasons for thinking that Bacon was a late medieval alchemist. Further, from what we know of alchemy in general, there is every reason to connect alchemy with Hermetic thought. Alchemy was based on the Pythagorean concept of the unity and harmony of all things. To repeat the central statement of the Hermetic *Emerald Table*: "That which is above is like that which is below: and that which is below is like that which is above to accomplish the miracles of one thing."[122] That being the case, it is hardly coincidental that Bacon's explanations of chemical and physical phenomena are often drawn from mystical theology. This may also explain why the publisher Macé Bonhomme would include within a single volume:

> Roger Bachon, *De L'Admirable, Pouvoir et Puissance de l'art et de nature, ou est traieté de la piere philosophale; Le Miroir D'Alquimie, de Rogier Bacon philosophe très-excellent; La Table des Meraude D'Hermes Trimegiste, père des philosophes;* and *Petit Commentaire de L'Hortulain Philosophe Dict des Jardins Maritimes, sus la table à Esmeraude d'Hermes Trimegiste.*[123]

The *Speculum Alchemiae* (Nürnberg, 1541, published in English as *The Mirror of Alchimy* in 1597) has been variously positively identified as a product of Bacon's hand and as having been "attributed" to him. If, indeed, we take the work to represent Bacon's thought, it gives us the double advantage of portraying Bacon as a convinced alchemist and of affording us insight into his alchemistic experiments.[124]

> *Alchimy* [it points out] is a Science, teaching how to transforme any kind of mettall into another teaching how to make and compound a certaine medicine, which is called *Elixir*, the which when it is cast upon mettalls or imperfect bodies, doth fully perfect them in the verie projection.[125]

The alchemy of the time was based on the belief that all metals were generated in the womb of nature by different combinations of sulphur and mercury. The alchemists set up their laboratories to duplicate nature's processes, yet at a considerably faster pace than was possible in nature itself. The aim of such experimentation was to combine "pure sulphur" with "pure mercury" in exactly the correct proportions that would produce pure gold, the perfect metal. Even more important, it was an attempt to distill the *fifth eternal essence*, the *quinta essentia*, from any compound whatsoever in the belief that the application of this "philosopher's stone" or elixir to the body would be effective in healing and if administered to the soul it would bring about regeneration.[126]

Although the ideas of alchemy arise from an odd mixture of neo-Platonic theology, Hermetic philosophy, mathematics and experimental method, this does not discount their value *in toto*. False as was its premise, Bacon's requirement that laboratory procedures emulate nature both in "simplicity" and in the "process of concoction" resembles modern scientific rigor, if only at a rather elementary level. Because what was needed to produce gold was not natural sulphur and mercury but rather their purified forms not found in nature, procedures involving speculation, observation, experiment and control set out to bring about results that would surpass nature's own.

> . . . we may find out a thing or some body of as cleane, or rather more cleane Sulphur & Argent-vive, on which nature hath wrought little or nothing at all, which with our artificiall fire, & experience of our art, we are able to bring unto his due concoction, mundification, colour and fixation, continuing our ingenious labour upon it.[127]

The alchemist was instructed to control the processes carefully and note the change in the materials when, at different temperatures, they changed from black to white or from white to red. It was, after all, the *individual units* of matter which were important, insisted Bacon. "Matter is not *unam numero*", in contradistinction to accepted doctrine,[128] for there are different kinds of matter. "An ass is not different from a horse by its form alone but it is a difference of specific matter".[129] Following Grosseteste's teaching the units of matter were for Bacon subject to quantification and control by mathematics by which alone certainty is given (*in sola mathematica est certitudo sine dubitatione*).[130]

Hence, alchemist though he was or because he was an alchemist, Bacon became an experimenter. His observations, his concentration on the individual units of matter, his use of mathematics and his noting of the results of his experiments for further use along with his experiments in optics all point to modern science in the process of evolution.¹³¹

It is important to note that for Bacon, experimentation was not simply a theoretical matter confined to the laboratory. His vision of universal science and of applied science or "technology", as we would designate it, is something we would expect from the seventeenth century. Indeed, it closely resembles the ideas of Francis Bacon.

For Roger Bacon science has three "dignities". The first is *scientia experimentalis*, best translated, perhaps, as "experiential" rather than as "experimental" science. Verification of the knowledge that one receives by revelation about nature is established on the ground of experience. We have illustrated this in the case of the rainbow. The first stage in the process of verification demanded *credulity*, the taking for granted of hypotheses. These were either eliminated or confirmed by comparing them with one another and by subjecting them to mathematics. Anticipating what William of Ockham was to call "evident intuitive knowledge", Bacon, on the one hand, warned against knowledge of natural things by revelation and deduction. On the other hand he cautioned against the imperfect experiences of the operative sciences. This would seem to imply the use of intuition as a basis for setting up hypotheses. These, in turn, were to be tested by means of experience and the kind of experiment then available.¹³²

The second "dignity" of science, according to Bacon's scheme, was *synthesis*. This is a compendium of scientific knowledge stated in a form which was true, well-chosen, systematic, concise, experientially-based and as perfect as possible. All sciences were to be included so that their results could be coordinated with a view to their mutual enlightenment.¹³³

The third "dignity" is what Bacon referred to as *the practical possibilities of science*, or what we call applied science and "technology". Here Bacon exhibits both what we recognize as the possibilities of "scientific-technological" knowledge along with what one might consider to be unwarranted speculation. He envisioned not only applying mathematics to nature by way of alchemy to make gold, but he foresaw also flying machines, powered boats, bridges suspended in the air, arrangements of mirrors to detect armies at a distance and of "burning glasses" large enough to destroy them.[134]

Bacon knew of the invention of explosive powder and apparently experienced it in the form of what we would call "firecrackers". The sound of that "toy for children", he wrote,"exceeds the roar of sharp thunder, and the flash exceeds the greatest brilliancy of lightning accompanying the thunder".[135] According to H. W. L. Hime's analysis, Bacon put down his first formula for the mixture of saltpeter, charcoal and sulphur only in cryptogram sensing, no doubt, its potential danger. The potential became actual when, around 1330, the German Franciscan Berthold Schwarz (fl. 14th cent.) invented a tubular device that could utilize the mixture to propel a projectile. With that, Bacon's "explosive powder" became known as "gun powder".[136] The device, as it turned out, was much more efficient at destroying armies than were Bacon's mirror arrangements, but in light of laser technology, and the "star wars projects" Bacon, perhaps unfortunately, may turn out to have been right on this count as well.

Through all of this Bacon remained a Franciscan monk who was certain of the correctness of the Christian faith. It should not surprise us therefore that he was convinced of the interrelation between faith and science. Nonetheless, it is important to note the correlation that he makes between the two, especially as we now are again beginning to appreciate the relationships and inter-impingements between our thought about God and our thought about the world.

At a basic level, Bacon emphasized that it behooved the Christian to study science since knowledge is part of the wonderful works of God. Further, Bacon was convinced, that science enables us to plot out the human future. It has a contribution to make to the last end of man.[137] In addition, Bacon claimed that the Christian was in the privileged position of having access to truth through grace. For Bacon God was behind scientific knowledge. The discoveries of science depended on God's divine revelation. Therefore, the Christian was much more likely than the heathen to gain scientific knowledge.[138]

This is not to say that the heathen had been bereft of scientific knowledge altogether. Bacon was convinced that due to the influence of the Hermetic tradition a universal science had been revealed fully to the patriarchs and prophets, partially revealed to Aristotle and slightly more fully to the Arabs.[139] However, the Judeo-Christian West or the "Latin world", as Bacon preferred to call it, had a special opportunity. As S. C. Easton has explained with reference to Bacon's position, the truths of Christianity had been revealed to this world. "Through God's grace a Latin Christian, if he lived a suitable life, might be vouchsafed the opportunity of completing it [a universal science] once more."[140]

For Roger Bacon, then, as for Francis Bacon, who was to become the spokesman for the implications of the relationship of faith to natural science some 400 years later on, faith, science, and the Christian life complemented one another. There seems to be no kinship between Roger and Francis Bacon, and no evidence that Francis ever read Roger Bacon's works. The agreement between the two, however, is startling.

Roger Bacon's opposition to the Aristotelian deductive logic, whereby problems were solved through logical syllogism, his attempts at experimentation and his classification of science, could serve as types for Francis Bacon's antitypes. Likewise, Roger Bacon's realization that in scientific activity one does not have to duplicate nature. Rather, "to beat it at its own game", as it were, was a perspective that Francis Bacon developed in detail.

Roger Bacon's linking of faith with scientific endeavour, and not least his interest in Hermetic ideas, are astoundingly similar to the emphases of Francis Bacon's *De Augmentis Scientiarum* and *Novum Organum*. Indeed, if we remember that Roger Bacon insisted on the necessity of mathematics as Francis Bacon did not, we might even say that, from a scientific standpoint, Roger Bacon was perhaps more advanced.[141] It is in this light that Adamson writes that Roger

Bacon "was doing in the thirteenth century what Francis Bacon tried to do at the beginning of the seventeenth".[142]

The Inductive Method of Duns Scotus
If late medieval science began at Oxford with Grosseteste and was continued and expanded by Roger Bacon, this "Oxford movement" was brought to something of a conclusion, especially as regarding its philosophical implications, by John Duns Scotus (c. 1265-1308). Duns Scotus was a philosopher rather than a scientist. Although he did not contribute to natural science as a natural scientist, his thought is quite indispensable to our understanding of the development of science in the West. As Torrance has said, "certain features in his conception of intuitive and abstractive knowledge . . . helped to create ferment and introduce the change that eventually came to fruition in the sixteenth century".[143]

We have already observed that as early as Philoponos, knowledge of the world began to be based on attentiveness to particular objects. The nature of the universe began to be known by way of observation rather than by way of deductive logic. This movement toward empiricism was advanced by both Grosseteste and Roger Bacon. It moved from practice to theory and led to the *raison d'être* of the development of a distinctive scientific method in Duns Scotus. Consequently his thought is basic to modern epistemology in general.

Like Grosseteste and Bacon, Duns Scotus insisted upon correlating particulars and investigating these for instances of uniformity and economy. Thus he hoped to intuit realities in accordance with their specific nature and particularity.[144] As Crombie shows, Duns Scotus and later William of Ockham both used Grosseteste's commentary on Aristotle's *Posterior Analytics* wherein Aristotle introduces his form of deductive thought. It is, therefore, not surprising to find Duns Scotus in agreement with his two Franciscan predecessors, Grosseteste and Bacon. All agreed that the way of induction was the way for epistemology.

In contrast to Aristotle's deductive method, induction

prescribed reasoning from the particular to the general rather than vice versa. On this point Crombie states that Duns Scotus "adhered entirely to the Aristotelian theory that knowledge of universals was reached only by abstraction from experienced particulars".[145] However, Torrance argues, that although Duns Scotus like Aquinas indeed held that scientific knowledge is concerned with the process of abstracting and of considering universals, he concentrated on the "specific existence of singular things in their individuations". Thus, to follow Torrance, the fact that for Duns Scotus the essence of things was to be found in singularities or particulars and the objective reference to universal concepts was rooted in singularities or particulars shows that Duns Scotus "made intuitive knowledge primary and abstractive knowledge secondary".[146]

It is true that Aristotle had long since realized that knowledge of things in general, "universals", could be obtained only by examination of the particulars of a selected sample of objects "proving the universal from the self-evident nature of the particular".[147] As Duns Scotus explained, this was possible because the principle of uniformity (inherited from Grosseteste) insured that objects were possessed of the same qualities as the sample under scrutiny.[148] There was, therefore, something of the universal in the particular and for Duns Scotus it was in the examination of things in themselves (the particulars) that both knowledge of them and the causes which are of their nature were to be found.[149]

This emphasis on the particular, or what may be called "the movement toward empiricism", had the effect of cutting at the roots of both Augustinianism and Aristotelianism at three levels. First, the Augustinian doctrine that particular things are to be considered only as representations of higher universal truth was put into question. Second, the Aristotelian concept that particulars represented the "accidents" of a "higher" and more universal substance was undermined. Third, Aristotle's idea that knowledge is possible only through syllogistic deduction from higher authority, against which Bacon had protested so vehemently, was thoroughly undone.

Duns Scotus may well have overestimated "experience" when, in the so-called *Oxford Commentary*, he insisted that "one who knows by experience knows infallibly that things are thus, always thus, and thus in all".[150] Nevertheless, his emphasis on concrete particularity that led directly to his theory of causality is shown by the statement: "Whatever occurs as in a great many cases from some cause which is not free [*i.e.*, not the result of free will] is the natural effect of that cause",[151] opened the way for the kind of trust in the cause-effect nexus which is basic to experimental science. Causes are thus "learned" or rather "intuited" from the observation of a number of similar cases or from "a sample".[152] Cause and effect were not the result of the intervention of an outside agency, rather they constituted the essense of nature itself.

Further, in that Duns Scotus believed that nature was *uniform*, once discovered, the relations of cause and effect could be treated as matters of principle from which knowledge could be inferred even without direct experience. Here, then, a quite legitimate level of abstraction seems to have been reached. Duns Scotus used the lunar eclipse as an example of his method. Once it was known that the interposition of the earth between the sun and the moon would block the light of the sun from reaching the moon, it was not only easier to recognize the phenomenon when it occurred but it was possible to demonstrate the nature of an eclipse from fundamental principle without actually experiencing an eclipse.[153]

Duns Scotus himself was not known as an experimenter. Nevertheless, like Francis Bacon some three hundred years later, his ideas lay at the very foundation of experimental science. He understood both the necessity of gaining generalizations from particulars, and that once reached by induction, generalizations could be treated as principles for deductive purposes. We can recognize here the beginnings of the hypothetical-deductive method that is basic to the whole of modern experimental science. Duns Scotus' principle of economy was as old as the Pythagoreans and was inherited from his Franciscan predecessors.[154] Nevertheless his restatement of it was passed on to his successors and especially to William of Ockham who gave it its classical enunciation. In summing up the philosophical implications of the "Oxford School" of medieval science, he laid the epistemological basis for future developments.

On the one hand, the contributions of the "Oxford School" can now be seen as little less than amazing. Grosseteste has been appreciated especially by Crombie, who finds in him "the beginning of modern science". Redgrove has named Roger Bacon "the father of experimental science". On the other hand Torrance's appraisal of Duns Scotus has helped put the "Oxford School's" contribution in a broader perspective. Howe offers additional insight when he points out that, although Grosseteste and those associated with him began to develop proper scientific procedures on the basis of Aristotle, neither Aristotelian logic nor the mathematics of the Middle Ages was sufficient for systematic natural scientific procedures.

As we have seen, Grosseteste and Roger Bacon both learned from and, to a certain extent, were dependent on Aristotle. However they, as well as Duns Scotus, moved against and away from Aristotle's deductive method that, from Aquinas onward, had become foundational to medieval thought. Therefore, it became as much a hindrance as a help to the development of natural science. Aristotle was not easily put aside, however. His deductive logic and metaphysics, which were of a piece with his physics, remained the guiding principle behind the medieval church's theology and "science" which was, to a large extent, beholden to theology at this juncture.

An Assessment of Aristotelianism
Thus, in spite of our respect for the scientific work of Grosseteste and Bacon, or the thought of Duns Scotus who continued the Oxford tradition in his philosophical works, the contribution they made ought perhaps be understood as preliminary to natural science rather than as natural science in the modern sense. Such judgments are arbitrary,

of course. They are arguable and in many cases their antecedents are difficult to assign.

If, as Walter Böhm has pointed out, many if not most of the concepts with which the scientists and natural philosophers Alhazen, Grosseteste, Bacon, Duns Scotus, Ockham, Buridan, Galileo and Kepler operated were anticipated and in many instances explicated by Philoponos then we are often dealing with rediscoveries rather than discoveries.[155] The exercise remains valuable nevertheless, for while Philoponos, lived in the open atmosphere of Platonically and neo-Platonically-influenced Alexandria, Roger Bacon, Duns Scotus and William of Ockham battled directly with a more dogmatic Aristotelianism. In the twelfth century Aristotelianism was the system by which the world was learning to live and by which it began to identify both scientific and Christian reality. In the thirteenth century Aristotle slowly but surely began to dominate the late Middle Ages. So trenchant was his hold on the medieval mind that non-Aristotelian thought was simply considered heterodoxy.

As seen from the perspective of history, Aristotle was questioned by what may best be called "experiential science" from John Philoponos onward. In fact, prior to Philoponos, early Christian writers from the post-apostolic period — Clement of Rome (fl. c.96), Clement of Alexandria (c.150 - c.212), Origen (c.185 - c.254), and especially Basil the Great (c.329 - 379) — had already set the stage for questioning the Aristotelian conceptuality of the eternity of the world, the division of both the heavens and the earth, and both the corruptibility and non-eternality of the "supra-lunar" and "sub-lunar" spheres. From Origen in the early third century to the twelfth century author of περὶ διδάξεων *sive elementorum philosophiae* (*Of Philosophical Elements*), which has been variously attributed to the Venerable Bede (c.673 - c.735) or to William of Conches (c.1084 - c.1154), Christians argued that the waters above the firmament in the Genesis account of creation were of the same substance, be they liquid, solid, or vapourous, as water on earth. Likewise, the fire of the stars was to be compared with fire on earth. Even a circle, Basil insisted, which though once created had no beginning and no end, if drawn had to be begun at a certain point.[156] Augustine, too, whose neo-Platonism provoked in him little interest in creation as such, nevertheless insisted on biblical grounds on the createdness of all things visible and invisible.[157] Augustine remained sufficiently beholden to Platonic and Neoplatonic thought, however, to allow that the heavenly waters were immortal and clear of worldly corruption.[158]

Philoponos, of course, challenged Aristotle both from the point of view of the non-circularity of the heavenly movements and on the basis of the comparison of the fire of the heavens with that on earth. Consistent with Aristotle's physics (but not with his astronomy) in that which could be seen was supposed to be tangible, and that which was tangible was considered to be corruptible, Philoponos was certain that the heavenly bodies must be composed of elements which were earthly rather than of an ethereal, non-material and eternal element, a *quinta essentia*.[159] Likewise, the unknown author of the treatise, *Of Philosophical Elements*, thought of the heavenly bodies as consisting of material. He speculated that the sun and the stars were largely composed of fire but consisted of other elements as well.[160]

The implications of the doctrine of *creatio ex nihilo*, as based on the transcendence of God over the totality of creation, relegated the heavens, earth and time, and the phenomena of impetus to finitude. Thus understood this doctrine began to have implications for the understanding of created nature from the early Christian centuries onward. So far ranging was the conviction of the biblical understanding, whereby God was different from but responsible for the world, that in a real sense the biblically-inspired view of creation prevented Aristotelianism from becoming a completely monolithic system in the West.

Considering the fact that both Grosseteste and Roger Bacon had already set much of Aristotle aside with their proto-empiricism, one may well conjecture that had Averroës and Aquinas been scientists as well as philosophers, the church and the West and the culture they dominated might never have had to endure the straitjacket of Thomistic Aristotelianism. The opposite point of view, represented by Whitehead, is that modern science is an unconscious derivative of medieval theology.[161] As we have seen, however, that argument is at best a half truth. Aristotelian deductivism and with it medieval theology had to be broken for experimental science to begin.

Over against the usual contention that it was Aristotle alone who opened the West to Greek science, we now know that Philoponos' writings, including his critique of Aristotle were known to the Arab thinkers Alghazali, Avicenna, Averroës, and Alhazen. Through them as well as by way of Simplicius, Philoponos was also known to Albert the Great, Thomas Aquinas and even to Galileo. The opinion that the West was totally dependent upon Aristotle as transmitted by the Arab scholars for its scientific interest would seem, therefore, to be in need of qualification. Philoponos' commentary on Aristotle's *Physics* not only presented Aristotle's arguments but refuted them in the process. Grosseteste did the same. In addition Platonic thought that had been kept alive in the West first by Augustinianism and then by neo-Platonic, neo-Pythagorean Hermeticism remained a power to be reckoned with.[162]

Even though some of Philoponos' thought was known to the West in the late Middle Ages, the "scientific" methodologies of Grosseteste and Bacon along with the epistemological arguments of Duns Scotus were not superfluous. It is no coincidence that the positive contributions of Philoponos, Grosseteste, Bacon and Duns Scotus to the development of science in the West were accompanied by their trenchant attacks on Aristotle as he was interpreted by the Averroists. From a twentieth century point of view, which is as dominated by the concepts of natural science as the late medieval period was by Aristotelian philosophy, it might seem that the Aristotelian understanding of things should have been discredited to the point that it would have been forgotten. This is, however, both to overestimate the persuasive power of the scientific development of the time and totally to underestimate the hegemony that ecclesiastical Aristotelianism had over the medieval mind. Looking back with regard to the development of the scientific mind-set we tend to see Aristotelian thought in the late middle ages as crippled and struggling to maintain its balance. In reality, however, it remained alive and well enough to hold sway within the church and culture throughout the medieval period and beyond. Challenged though it was, it kept experimental science at bay for yet another three hundred years.

Ockham's Empirical Logic
In many ways the "hinge of history" that connected the medieval world and that of the Reformation was William of Ockham (c.1284 - c.1349). Like Duns Scotus, Ockham was not an experimental scientist. Nevertheless, his logic cut so deeply at the roots of Aristotelian deductive thinking in the late Middle Ages that it survived only by the dint of ecclesiastical and political power. As has been said, Ockham's interest and talents were not primarily "scientific," at least not "natural scientific". However, the fact that he used Roger Bacon's *scientia experimentalis* as a basis for his own *scientia intuitiva*, indicates that he developed his philosophical ideas in direct relationship to the most advanced scientific conceptualities of the time.[163]

In Ockham's hands the philosophical and epistemological implications of the kind of thought structures we have noted in Philoponos, Grosseteste, Roger Bacon, and Duns Scotus take an even more incisive character than was the case with their originators. Ockham picks up where his predecessors left off and transforms their insights through the formulation of a logic with universal applicability. The result was two-fold. First, this move was in itself a fundamental attack on medieval metaphysics. Second and equally important, he initiated a separation between the question of knowledge concerning faith, on the one hand, and that of knowledge as related to natural science, on the other. With these two Ockham provides the epistemological foundation for the transition of thought from the late Middle Ages to the Renaissance. From there its role continued through the Protestant Reformation in the sixteenth century to the flowering of natural science in the seventeenth.

In the first instance, Ockham built on Duns Scotus' stance that thought must concentrate upon the distinctive nature of particulars, for it is only there that evidence of reality is available. The method, like that of Duns Scotus and Bacon, was inductive instead of deductive. For Ockham, it was induction with a vengeance. He set up his epistemological rationale by making a primary distinction between the science of real entities (*scientia realis*), and the science of logical entities (*scientia rationalis*). Whereas *scientia realis* had to do with the knowing of things by empirical experience, *scientia rationalis* was concerned with the recognition of logical concepts by way of reflection. Although Ockham recognized the possibility of giving each rational concept a name (*nomen*, hence "nominalism"), such were "rational realities" and differed fundamentally from "empirical entities". Hence, Ockham contrasts the method of *scientia rationalis* with that of *scientia realis* by showing that it is brought about by intuitive knowledge (*notitia intuitiva*).[164]

Reality, for Ockham, consisted of individual realities only. He considered the generalities by which the aggregates of things were identified simply as logical constructs to which names were assigned. There was only "this house" and "that house". The name, "houses" was simply an abstraction, only a name (*nomen*). Ockham's emphasis is perhaps best referred to as "particularism".[165]

When Ockham's thought is taken seriously, both Augustinian sacramentalism, wherein the material was but a sign or signification of the spiritual, and the Aristotelian-Thomist conceptuality of substance-accidents, wherein the particular is only the accidental manifestation of a higher general substance, became untenable. Of equal importance to the breakdown of the Aristotelian system of deductive logic is, as Crombie has suggested, Ockham's radical rejection of absolute causation. The result was that the observed regularities of the world came to be understood as mere regularities of fact. The laws expressing these regularities were thus demoted to being at best possibilities. At worst, they became conventional devices for correlation and calculation.[166]

In current Ockham critique a comparison with Hume has played an important role.[167] There is little doubt that the implications of Ockham's stance have something of the implications of David Hume's (1711-1776) scepticism toward the post-Newtonian concept of causation. One might even point to a coincidence of the two positions. In turning against the rationalistic understanding of causality in the Aristotelian sense, Ockham freed the mind from the hegemony of deductive logic that argued from the truth of first principles to inexorable conclusions and made way for reasoning intuitively from the particular to the general. Similarly, Hume's objection to Newtonian causality opened post-seventeenth century science to the possibilities of relativism. Hume thus awoke Kant from his dogmatic slumbers. Kant, in order to re-establish the assured validity of the scientific intercourse, moved causality from the world to the mind and thereby laid the foundation of his transcendental idealism.

Ockham's rejection of causation as an absolute category was directed at the *first* and *final* causes that for Aristotle were the be all and end all of any investigation. He was, however, quite willing to posit causal relationships as intuited from an aggregate of particulars. In that Ockham, along with Roger Bacon and Duns Scotus, accepted the principle of the uniformity of nature, Ockham not only accepts causation at the level of like particulars but he builds the basis for investigation. "All individuals of the same kind are so made as to have effects of the same kind on an object of the same kind in the same circumstances."[168] In contrast to Aristotle, Ockham insisted that causal connection must be conceived and established empirically, that is, in immediate connection with individual realities under observation. Probable causes were first intuited from the connection between one particular and another. Following his law of parsimony Ockham eliminated any "so-called causes" for which no observable effects could be established.

In this way Ockham moved toward *empiricism* and away

from *rationalism*. He does not deny causation as such, but is careful not to define any probable cause in absolute terms. Thus, it is quite clear from the epistemological structure that results from the way he formulated rules for causal connections on the basis of "agreement and difference". For an "immediate cause" (and all causes for Ockham are "immediate") to be present there must be agreement between a cause and its effect. If difference is observed, the cause is not present. "When it [the cause] is present, the effect follows and when not present, all other conditions being the same, the effect does not follow".[169] Hence, "usually it is not possible to know a singular contingent proposition clearly without many apprehensions of single instances".[170] Ockham's attempt was to eliminate what we would call false hypotheses in order, "to arrive at *as certain as possible a knowledge* of principles from which demonstration could follow".[171]

The process follows the inductive method of Roger Bacon and is directly opposed to Aristotle as he was transmitted to the West by Averroës. Principles cannot be demonstrated by syllogism from better known propositions. They are arrived at only by intuitive knowledge.[172] A simple diagnostic procedure is as follows:

> Since he observed that after eating such herbs the fevered person was cured and he removed all other causes of his recovery, he knew *evidently* that this herb was the cause of recovery, and then he has experimental knowledge of a particular connection.[173]

Cause, then, like substance, is a matter of immediate relationship to the reality of the particular. "Final causes" are ruled out. According to Ockham, the special characteristic of a final cause is that it is able to cause when it does not exist.[174] Final causes, like universals, are "not real but metaphorical".[175]

In giving formal expression to the practice of Grosseteste and especially of Roger Bacon, he set up a system wherein "the only reliable knowledge of contingent, natural events was that acquired by observation". Beginning with a number of instances, the procedure was first to establish principles by means of induction and correlate these with observed facts by means of logic and mathematics.[176] In this way Ockham's thought predisposed the "scientists" or "natural philosophers" as they were then known, to seek knowledge of nature by means of experiment.[177]

Such was Ockham's "scientific method" based on induction rather than deduction and on intuition rather than abstraction. The procedure was regulated by the "principle of economy" according to which the simplest explanation was to be preferred. The principle, which was as old as the Pythagoreans and which was used by Ptolemy to decide on the best system for explaining the rotation of heavenly bodies, was apparently transferred to the West by Averroës.[178] Ockham, like Grosseteste and Bacon, considered it not only a principle of method but "a principle of nature". The "principle of economy" simply meant that nature operated "in the shortest way possible".[179] Ockham's formula was: "a plurality is not to be asserted without necessity".[180]

Although the principle of economy was used by the Pythagoreans and was adopted by Plato, Aristotle and Ptolemy, it is quite evident that, as such, it is no guarantee of truth. Nevertheless, the concept of economy, like those of simplicity, symmetry, homogeneity, isotropy, beauty and elegance, most if not all of which the Pythagoreans had applied to the heavens, are the non-provable but indispensable "articles of faith" that have governed decisions from the beginnings of natural science to the present.[181]

In addition to substituting his logic of induction from particulars for Aristotle's concept of gaining knowledge by a process of deduction from first principles, Ockham augmented his attack on Aristotelian metaphysics and physics by promoting three hypotheses. The first and third of these hypotheses appear to recall the teaching of Philoponos although they are not necessarily dependent upon his teaching. The first hypothesis asserted that heavenly and earthly bodies were composed of the same kind of material. This had been emphasized by Christian thinkers in one way or another from the first century onward. Ockham's reiteration of it indicates, however, that the consequences of the doctrine of *creatio ex nihilo*, which entailed both the transcendence of God over creation and the unity of all creation under God, were easily ignored. Ockham's second hypothesis which asserted that any part or piece of material tends to move toward its whole or it strives to fill a vacuum is again a stress on the isomorphic character of nature.[182]

The third hypothesis had to do with motion, the theory of "impetus" which some four centuries later was to play an absolutely essential role in Galileo and Newton. Rather than following Aristotelian physics and distinguishing between motion and the thing that moves, Ockham asserted that that which moves and motion were to be considered identical.[183] Thus after leaving the mechanism which set it in motion (*e.g.*, a catapult), a projectile is its own mover (*ipsum movens . . . est ipsum motum secundum se*), and it is impossible to distinguish between what moves and what is moved (*movens et motum est penitus indistinctum*).[184] This had the result that motion itself, the most important aspect of nature which Greek physics was never able to explain, was made accountable by the method of empiricism.

Ockham combines his empirical method, his principle of economy, and his theory of impetus in a single statement:

> It is evident that given that one body is in one place and later in another place thus proceeding without any rest and any other intermediate thing save the body itself and the agent itself which moves, we have true local motion. It is, therefore, futile to postulate other such things.[185]

The implications of this statement are obvious. While it is true that, for Ockham, relations were "concepts formed by the mind", a relation was not an abstraction but a reality that connected verifiable particulars. A relation "had no objective reality apart from the individual perceptible things between which the relation was found".[186] This, together with Ockham's conviction that we have observational experience not of "substance" but only of "attributes" (to follow the prevailing Aristotelian terminology), conflicted sharply with the Aristotelian view that the cosmos was characterized by "an objective principle of order according to which its constituent substances were arranged".[187]

Hence, as far as contingent, *i.e.*, created, things were

concerned, Aristotle's first and final causes were set aside. Aristotle's God who impregnates contingent reality with final causes was made redundant. Creation operated according to its own inherent laws and these were to be found by investigating the individual actualities of nature. In this way the interrelations between things were spelled out.

In consideration of his predecessors, Philoponos, Grosseteste, and Roger Bacon as well as Duns Scotus, Ockham turns out to be less original than was once supposed. Nevertheless we may take seriously Crombie's statement that:

> The effect of Ockham's attack on contemporary physics and metaphysics was to destroy belief in most of the principles on which the 13th-century system of physics was based. In particular, he attacked the Aristotelian categories of 'relation' and 'substance' and the notion of causation.[188]

Ockham did not apply his scientific approach to theology, however, at least not directly. Rather, formally at any rate, he continued to insist that the only *true knowledge* was given by revelation. On the one hand, then, he advocated that science or natural philosophy was dependent on sensory experience gained by the observation of particulars. On the other hand, he held that theological understanding was dependent upon divine revelation alone. The effect was to drive a wedge between two incompatible elements in the Thomistic medieval synthesis. Knowledge of revelation according to the perspective of Christian faith was one thing. Knowledge of the world according to the perspective of natural philosophy was quite another. Ockham's intention was not to oppose faith to philosophy nor revealed knowledge to natural knowledge as might be supposed at first reading and as was the case with Plato. Rather, as Torrance points out, "He showed that they involved two very different approaches and therefore quite different notions of *scientia*".[189] He recognized, one might say, "two realms of discourse", two realms that Aristotle had amalgamated into one.

Ockham differentiated between revelation and reason, and he also distinguished God from nature. God may be known only through revelation and "infused faith", but nature can be known only through observation and reason. The revolutionary character of Ockham's thought at this point is apparent when compared to that of Thomas Acquinas with which it stands in sharp contrast. Whereas the Thomist concept that divine forms interpenetrate nature led to the conclusion that knowledge of God could be read directly from nature through reason, Ockham's concept of nature revealed only nature. In contrast to Aristotle, who filled nature with divine first and final causes and made the knowledge of nature and knowledge of God so inseparable that knowledge of the one was coterminous with knowledge of the other, Ockham returned to the biblical doctrine of God. Accordingly God was the transcendent Creator who created the world *ex nihilo*. In this way, too, his perceptions represent a direct break with the prevailing Aristotelian thought as incorporated in medieval theology.[190]

Parallel to his distinction between the way of knowing revealed and divine truth, on the one hand, and knowing contingent reality, on the other, was Ockham's differentiation between the realm of the church and the realm of the world. This in turn led to the extremely crucial difference which Ockham made between spiritual and temporal power that was indispensable for the development of science from the fourteenth to the seventeenth centuries. From the perspective of the Scriptures which for him were normative, Ockham insisted that the papacy's claim to be invested with power over the "heavenly" and "earthly imperium" was erroneous. "Just as no one holds the heavenly imperium from the pope in fief, so also no one holds the earthly imperium in fief from him".[191]

With both the structures of his logic and his exegetical conclusions, Ockham challenged the monolithic authority of the church, reinforced with the absolute structures of Aristotelian-Thomist thought over both mind and society at its roots. Two eminently important results followed. First, the dominance of the church over secular life was called into question. Second, the hegemony of theological thought over scientific thought as a whole and over the logic of the sciences in particular was shown to be without foundation.

In our time we have learned to regret the sharp distinction drawn by Ockhamist thought between faith and reason. In the development of philosophical thought, it was this kind of fundamental dualism as emphasized by René Descartes (1596-1650) that led to the dichotomies in thought and life that characterized the epistemological malaise in which we find ourselves today. In the fourteenth century, however, as indeed at the beginning of the seventeenth, the concentration of power in the church and the influence of the church over culture in general made a separation of reason from "faith" quite necessary for the freedom of scientific thought.

The *diastasis* created between the two served to compromise the monolithic thinking that followed the pattern of Aristotle's deductive logic and was instrumental in breaking its stranglehold on the medieval mind. Ockham's thought gave a *raison d'être* to the investigation of material nature leading to the understanding of nature as nature rather than conceiving of it as a mere accident of an eternal substance, whose essence was to be deduced syllogistically from preconceived and unquestioned first principles.

Thus Ockham's "two-realm theory of epistemology" was initially of immense import. However, when its impact was continued "beyond its time" it had negative consequences. In its own time, however, Ockham's insights into the nature of analytical and critical knowledge that summed up the anti-Aristotelian elements in the development of thought from Philoponos onward, were quite indispensable to the development of science. They stood over against the prevailing Aristotelian deductive system that Thomas Acquinas had incorporated into his medieval synthesis wherein things were identified according to the universal categories to which they supposedly belonged and whereby phenomena were considered simply as manifestations of a chain of eternal patterns.

In concentrating upon particulars as particulars Ockham prepared the way for the new structures of thought on which the whole of modern western civilization was to be built. As Torrance has put it, it was Ockham's insistence that knowledge of contingent things is possible only through the experience of them as individual and singular which opened the way for the "development of an empirical habit of mind".[192] Individual objects were thus acknowledged to have identity and worth in themselves and not only in a secondary way as "accidents" of a more important and overarching "substance", as Aristotle and Aristotelian Thomism would have it. By studying the object in conjunction with other objects, by intuiting the relation between the objects and finally by applying that relation to "like" objects, the world began to be understood as an inter-related assemblage of particulars that could be understood according to the intrinsic relationships of the particulars themselves. With that the pathway to experimental science was opened.

The Fall of Aristotle
Ockham's attack on the Aristotelian notion of direct causation shows how far-reaching his "empirical way" of looking at things turned out to be. Whereas in his deductive mood Aristotle explained events in nature by tracing their immediate causes logically back to the first cause, for Ockham, causes had to be conceived in relationship to sensory experience. This "empirical habit" taken along with "Ockham's razor" showed that nature could be understood in relation to itself without recourse to the elaborate metaphysics of Aristotle as an explaination of the very basis of reality. In following the "empirical habit of mind", science began to be built upon a method that demanded direct investigation of empirical "facts and particulars" and an elemental albeit the "simplest" possible explanation of the natural connections between these "facts and particulars".

The effect of Ockham's anti-Aristotelian conceptions can be seen in Jean Buridan (c.1300 - c.1358), rector of the University of Paris and a student of Ockham's. He applied Ockham's logic to cosmology in a way reminiscent of Philoponos.[193] By concentrating on the particular and favoring the simplest of possible explanations, Buridan found Aristotle's divine intervening intelligences superfluous. In that God had provided the heavenly bodies with their impetus at creation, they had no need of intelligences that, according to the metaphysics of Aristotle, were supposed to be responsible for continuing their movement.

Hence, in an interesting statement which both puts Ockham's ideas to work and reflects the most advanced science of the time, Buridan interwove scientific reason with theological apologia to challenge, however tentatively, the ecclesiastical establishment of his day.

> One does not understand from the Bible that there are intelligences whose task it is to give to the heavenly circles their characteristic movement. From this one can gather that there is no necessity for such. One could rather say that God, as he created the world, imparted to each heavenly circle a movement according to his pleasure. He impressed upon each one the impetus which has moved it ever since in such a way that God does not have to move these circles anymore except in the general sense in which he bestows on all activity his constant assistance. That is also the reason why on the seventh day he could rest from all the works which he had made because he now allowed the things he had created to practise their activities and influences themselves in relationship to one another. The impetus that God impressed upon the heavenly bodies will, in the span of time, neither be weakened nor exhausted because in these heavenly bodies there is no tendency whatsoever to alter movement and because there is no other resistance which could destroy or exhaust this impetus. I offer all of this not as certain knowledge. I want only to plead with the masters of theology to instruct me, how in their view these things could behave in any other way.[194]

Buridan's stance, of course, is not totally unique. As pointed out earlier, the theory of impetus can be traced to Philoponos in the sixth century A.D. The beginnings may, in fact, be seen in Hipparchus in the second century B.C., who held that the essence of an object itself determined both its nature and its mode.[195]

It was the nature of a stone, for instance, to fall faster than a feather. Thanks to Galileo we now know that this is not the case. When the idea that the behavior of objects was to be traced to natural causes first by Ockham and then by Buridan, it encouraged the concentration on nature as nature. Physics became primarily a matter of attending to the physical rather than to the metaphysical, the primary mode by which reality was to be investigated and known. Granted that observation remained, as we would say, theory-dependent, that theme remained a metaphysic beyond the physical that was taken for granted, i.e., that nature was uniform, that it revealed itself in the particular and by the simplest of possible explanations, etc. It was a metaphysic that allowed the data of observation to be understood within a new rational order. Rational order was as finite as were the objects themselves that were known in their finite relationships to other finite objects. Further, and of extreme importance for later developments, rational order was integrally related to the objects that revealed that order even as they were understood in relationship to it.[196]

According to the above investigation, it would seem quite clear that Greek and particularly Aristotelian thought and logic had both contributive and inhibiting effects upon the development of science. It contributed enormously to the opening of the western mind to the importance of the world and in recognizing the application of intentionality and logic (however intransigent) to the understanding and classification of the objects of nature. Aristotle's contribution would have been greater had his *Posterior Analytics* with its inductive procedures been valued to the same degree as were his *Prior Analytics*, his *Physics* and his *Metaphysics* wherein deductive logic is *de rigueur*.

Eventually, however, his thought fails, because he, like Plato before him, does not differentiate the finite from the infinite, the temporal from the eternal. Hence, the material is but an accident of the eternal and the divine substance it represents. Likewise, the first and final causes that interpenetrate all things and which are the basic *raison de être* of all existence are divine. Serious as is the study of nature, however, science deals in essence with chimera and not with things in their "thingness", nor with things as reality.

It is in this context that we again can understand that Whitehead's claim that Aristotle's conception of the "rationality of God and nature" was "a prerequisite for the development of modern scientific theory", is only partially true.[197] The very damaging aspect of Aristotle's thought, that Whitehead not only failed to see as harmful but which he adopted as the basis for his own thought, is his insistence that the divine and the infinite interpenetrate the finite and are responsible for its structure. Both Aristotle and later Whitehead derived the idea from Plato who had learned it from the Pythagoreans. So pervasive is the divine rationality that rather than being transcendent over nature, it interpenetrates and informs it. Here nature both reflects God and in certain respects may be understood to be interchangeable with him. Thus Plato's reference to the world as a "blessed God"[198] is no more coincidental with this understanding than is Whitehead's basic concept that understands God as concretizing himself in reality.

From another perspective the transcendental idealist, F. W. J. von Schelling (1775-1854) composed a magnificently detailed, consummately consistent philosophical system grounded upon the belief that the material world is the manifestation of the infinite under the exigencies of finitude. This beautifully developed system expressed but another phase in the history of the influence of Greek cosmology. Some insight into this phase is afforded us through the theology of Paul Tillich (1886-1965), whose thought is intricately intertwined with that of Schelling. The fact that Tillich's thought was heavily influenced by Schelling was both a great strength and eventually a great weakness. In the end he, like Plato, Aristotle, and the neo-Platonists (with whom Tillich has more in common than his footnotes indicate) fail to maintain the biblical understanding of God as transcendent over and separate from nature. They fail to grasp the significance of the doctrine that God stands in a contingent relationship to nature.[199]

The confusion stems from Plato's concept of creation. Rather than creating the world out of nothing, God forms what we may call the "pre-existent" into the "existent". "Out of disorder he brought order".[200] Aristotle's doctrine that divine causes interpenetrate nature is likewise an adaption of Plato's divine "world soul", the "never-ceasing and rational life" that is "infused everywhere from the center to the circumference of heaven".[201] This *deus sive natura* conception which seriously affected the medieval view of the world and conflicted with the Christian doctrine of God who created the world out of nothing (*i.e.*, created nature and its rationality in distinction from himself and his own rationality), had to be set aside before the world could become the world of nature and science could become the study of nature proper.[202]

John Philoponos, Robert Grosseteste, Roger Bacon, John Duns Scotus, William of Ockham and Jean Buridan, all helped clear the ground of the Aristotelian conceptions that at one and the same time prevented nature from being appreciated as nature and prevented God from being appreciated as God. In so doing they opened the way for the development of modern science and for the development of theology that led to the Protestant Reformation.

NOTES

1   Cf. Nebelsick, *Circles of God*, pp. 1 ff.
2   Ptolemy's *He Mathematike Syntaxis (The Mathematical Collection)*, known as the *Syntaxis,* the classic exposition of Greek mathematics and cosmology in thirteen volumes, is one of Greece's most prestigious scientific efforts. In the course of time the work became known as *Ho Megas Astronomos (The Great Astronomer)* to distinguish it from *The Little Astronomer,* a collection of works by other Greek mathematicians and astronomers. The Arabs who referred to Ptolemy's work as *Megiste (The Great Collection)* prefixed the name with their definite article, "al" and thus gave the publication the title, *Almagest,* by which it is still known. Cf. Ptolemy, *The Almagest,* tr. R. Catesby Taliaferro, Great Books of the Western World, Vol. 16 (Chicago: *Encyclopaedia Britannica,* 1952).
3   Ptolemy's geocentric system (or actually, "geofocused" system since the earth was set somewhat off centre of the eccentric orbits of the planets) consisted of eccentric deferents which were set in orbit around the sun. On the rim of each deferent was centred an epicycle (or small circle) which carried the planet in question on its circumference. The combined circular motion of the eccentric deferents and epicycles explained the apparent irregular motion of the planets around the stationary earth in terms of the regular motion of the movements involved. Ptolemy invented the "equant",

The Christian Critique of Aristotle

a mathematical point which was both off centre of each orbit and set somewhat off the earth as the point from which the proper regular and circular motion of each planet could be calculated.

4   Aristotle, *Physics* II.tr. P. H. Wicksteed and F. M. Cornford, Loeb Classical Library (London: Heinemann, 1934), ii, 193b. Cf. Nebelsick, *Circles of God*, p. 239, fn. 121.

5   For Aristotle, mathematics in differentiation from physics and astronomy was not concerned with the physical bodies as such but only with abstractions from their properties and motions. Aristotle, *Physics* I, LCL (London: Heinemann, 1929), II. ii, 193b.

6   Cf. Nebelsick, *Circles of God*, pp. 25 ff., 71 ff., 200 ff.

7   Cf. F. E. Peters, *Aristotle and the Arabs* (New York: New York University Press, 1968) and Seyyed Hossein Nasr, *An Introduction to Islamic Cosmological Doctrines* (Cambridge, MA: Harvard University Press, 1964).

8   Ibn Khaldun, *Muqaddimah* (tr. Rosenthal), III, 152, cited by Peters, *Aristotle and the Arabs*, p. 194. Cf. *ibid.*, esp. Ch. VI, "Philosophical Movements in Islam", pp. 135-183 and the section on Ibn Rushd, pp. 215-220.

9   Joseph Needham, *Science and Civilization in China*, History of Scientific Thought, Vol. II (Cambridge: University Press, 1956), p. 581.

10 Stephen Toulmin, *Human Understanding* I (Princeton: Princeton University, 1972), p. 219.

11 C. F. von Weizsäcker, *Die Einheit der Natur* (München: Hanser, 1972), p. 84. Translation mine.

12 Cf. "Introduction", above, fn. 1.

13 Both Albert the Great and Thomas Aquinas had argued for the doctrine of *creatio ex nihilo* as early as the thirteenth century as over against Aristotle's insistence on the eternality of the heavens. Since, however, neither questioned the composition or movements of the heavens as set forth in Aristotelian astronomy and physics, the objections were formal rather than material. Cf. Cyril Vollert's argument in the introduction of St. Thomas Aquinas, *On the Eternity of the World* (Milwaukee: Marquette University, 1964), pp. 11-13; Albert the Great, *Commentarii in Secundum Librum Sententiarum, Opera Omnia*, Vol. XXVII, ed. Borgnet (Paris: Vivès, 1895) Dist. I, Art. X, p. 28; Albert the Great, *Summae Theologiae Secunda Pars* (Quest. I-LXVII), *Opera Omnia*, Vol. XXXII, ed. Borgnet (Paris: Vivès, 1895) Tract I, Q. IV, Mem. II, Art. V, pp. 91-108; Thomas Aquinas, *In Aristotelis Libros de Caelo et Mundo* (Romae: Marietti, 1952), I, l. xix-xxii; Nebelsick, *Circles of God*, "Thomistic Aristotelianism", pp. 149 ff.

14 Samuel Sambursky, *The Physical World of Late Antiquity* (New York: Basic, 1962), p. 157.

15 *Ibid.*, pp. 166. Cf. Walter Böhm, *Johannes Philoponos* (München, Schöningh, 1967), p. 27.

16 Cf. Wolfgang Wieland, "Zur Raumtheorie des Johannes Philoponus", *Festschrift für Joseph Klein* (Göttingen, Vanderhoeck, 1967), pp. 114-

68 The Renaissance, the Reformation and the Rise of Science

17 135.
Jaki, *Relevance of Physics,* p. 416.
18 John Philoponos cited by Simplicius, *In de caelo comment.,* ed. J. L. Heiberg (Berlin, 1894), 32,2, cited by Sambursky, *Physical World of Late Antiquity,* p. 171.
19 Sambursky, *Physical World of Late Antiquity,* p. 171. Cf. Plato, *Timaeus, The Dialogues of Plato,* tr. B. Jowett, Vol. 2 (New York: Scribner, 1874) 36.
20 John Philoponos, *De opificio mundi,* ed. G. Reichardt (Leipzig, 1897) IV. 12 (184,26), translated by Sambursky, *Physical World of Late Antiquity,* p. 159. Cf. *ibid.,* p. 164.
21 Philoponos (cited by Simplicius), *In de caelo comment.,* 89,16, cited by Sambursky, *Physical World of Late Antiquity,* p. 160.
22 Sambursky, *Physical World of Late Antiquity,* p. 160. Cf. Nebelsick, *Circles of God,* pp. 109 f.
23 Sambursky, *Physical World of Late Antiquity,* p. 124. Cf. Böhm, *Philoponos,* p. 349.
24 Böhm, *Philoponos,* p. 370. Cf. below, on Galileo's Dialogues.
25 Böhm, *Philoponos,* "Wirkung des Johannes Philoponos auf die Nachwelt", pp. 337-386. Cf. Anneliese Maier, *Die Impetus-theorie der Scholastik,* 2nd ed. (Roma: Edizione di Storia, 1951).
26 Böhm, *Philoponos,* pp. 27 f.
27 Karl Jaspers, cited by Howe, *Mensch und Physik,* p. 34.
28 Crombie, *Grosseteste,* p. 11.
29 Howe, *Mensch und Physik,* p. 26.
30 We shall find, however, that Aristotle is at best ambiguous with regard to the "rationality of nature".
31 Whitehead, *Science and the Modern World* (New York: Free Press, 1969), p. 13.
32 Alfred Edward Taylor, *Platonism and Its Influence* (Boston: Marshall Jones, 1924), p. 23. Taylor backs up his statement with a reference to Etienne Gilson, *La Philosophie au Moyen Age,* II.46: "Oxford, où vont affluer les sciences nouvelles empruntée aux Arabes, recueillera et fera fructifier l'héritage de Chartres; on y restera fidèle au platonism augustinien, on y saura les langues et l'on y enseignera les mathématiques dont Paris se désintéressera."
33 A. C. Crombie, *Robert Grosseteste and the Origin of Experimental Science 1110-1700* (Oxford: Clarendon, 1953), p. 13.
34 Cf. *Die Philosophischen Werke des Robert Grosseteste,* ed. Ludwig Baur (Münster: Aschendorffsche, 1912), p. 24, fn. 1. Cf. Ludwig Baur, *Die Philosophie des Robert Grosseteste,* Band XVIII, Beiträge zur Geschichte der Philosophie des Mittelalters (Münster: Aschendorffsche, 1917).
35 Cf. *ibid.,* pp. 19 ff.
36 Plato, *The Republic, The Dialogues of Plato,* 4 vols., tr. B. Jowett, Vol. II (New York: Scribner, 1874), Bk. VI, 508-509.
37 St. Augustine, *De Genesi Ad Litteram, Oeuvres Complètes de Saint Augustin,* Vol. 7 (Paris: Vivès, 1873), Cap. V. 19-20, pp. 10 f.

38 Crombie, *Grosseteste*, p. 13.
39 *Ibid.*, p. 107, translating Robert Grosseteste, *De motu corporali et luce*, *Philosophischen Werke des Robert Grosseteste*, ed. Ludwig Baur, p. 92.
40 *St. Augustine's Confessions*, trans. William Watts, 2 vols. LCL (London: Heinemann, 1950, 1951), II, Liber XII, Cap. XVII, pp. 324-328.
41 "Lux ergo, quae est prima forma in materia prima creata, seipsam per seipsam undique infinities multiplicans et in omnem partem aequaliter porrigens, materiam, quam relinquere non potuit, secum distrahens in tantam molem, quanta est mundi machina, in principio temporis extendebat." Crombie, *Grosseteste*, p. 104, fn. 4.
42 *Ibid.*, p. 104.
43 Aristotle, *Posterior Analytics*, LCL(London: Heinemann, 1960), II. xix, 100b. It is noteworthy that Grosseteste's procedure sets aside Aristotle's use of universals whereby he reasoned from the undoubtable to the absolute and back again to necessity by syllogistic procedures. Aristotle, *Prior Analytics* LCL (London: Heinemann, 1938), I. ix-xv, 30b-33b.
44 Crombie, *Grosseteste*, p. 94 translating Grosseteste, *Commentarius in VIII Libros Physicorum Aristotelis* from MS Digby 220, f. 88v; Merton 295, f. 114v. The notations are from Aristotle, *Physics*, II. ii, 193b where Aristotle differentiates between mathematics, physics and astronomy and Aristotle, *Metaphysics* (Books X-XIV, LCL (London: Heinemann, 1962), XI. iii, 1061a-1061b.
45 Plato, *Timaeus*, Vol. 2, pp. 34-35 and 55-56.
46 Cf. Nebelsick, *Circles of God*, pp. 48 ff, 71 ff.
47 *Ibid.*, "Greek Theology and Greek Science", pp. 1-32 and Aristotle, *On the Heavens*, LCL (London: Heinemann, 1939) II. iv, 287a. The explanation underlies the whole Ptolemaic system. Cf. Ptolemy, *Almagest*, Bk. I, pp. 5-8.
48 Clement of Rome, *The Epistle of S. Clement to the Corinthians*, tr. J. B. Lightfoot, The Apostolic Fathers, Part I, Vol. II (London: Macmillan, 1890), v. 20, p. 282. Cf. Nebelsick, *Circles of God*, "Science Encounters the Christian Faith", pp. 88 ff.. For an insightful explanation of contingency, cf. T. F. Torrance, *Divine and Contingent Order* (New York: Oxford, 1981).
49 Grosseteste, *De lineis angulis et figuris seu de fractionibus et reflexionibus radiorum*, *Werke*, pp. 60-61.
50 Grosseteste, *De generatione stellarum*, *Werke*, p. 32.
51 *Ibid.*, pp. 33-36.
52 Grosseteste, *De sphaera*, *Werke*, p. 11. Cf. Grosseteste, *De lue*, pp. 55 f.
53 Ludwig Baur, *Die Philosophie des Robert Grosseteste* (Münster: Aschendorffsche, 1917), p. 171.
54 Grosseteste, *De generatione stellarum*, p. 33, lines 2-21.
55 *Ibid.*, p. 36, lines 3-11.
56 Grosseteste, *De finitate motus et temporis*, *Werke*, pp. 101-106.
57 Grosseteste, *De sphaera*, Ch. I, pp. 11-16. Cf. Aristotle, *On the Heavens*,

70 The Renaissance, the Reformation and the Rise of Science

II. iv; II. xiv.
58 Aristotle realized the difficulty and posited separate intelligences for each of the heavenly spheres to allow for the peculiarity of their motion. Aristotle, *On the Heavens*, II. xii, 292a-292b.
59 Grosseteste, *De motu supercaelestium, Werke*, pp. 92-100.
60 Aristotle, *Physics* II, VIII. i, 250-252b.
61 *Ibid.*, VIII. v, 256a-256b.
62 Grosseteste, *De motu supercaelestium*, pp. 92-93. Cf. Nebelsick, *Circles of God*, pp. 70 ff. and 126 ff. Interestingly enough, Aristotle too was not satisfied to attribute the complex motions of the planets to the prime mover alone but has fifty-five astral divinities to push the spheres around. Aristotle, *Metaphysics*, XII. viii, 1074a-1074b.
63 For a fuller discussion of Grosseteste's cosmological objections to Aristotle, cf., Nebelsick, *Circles of God*, pp. 126 ff.
64 The text is translated by Crombie, *Grossetste*, p. 96 from Grosseteste, *Commentarius in Aristotelis Analytica posteriora*, i.8, f.8r.
65 "Hoc principium philosophiae naturalis, scilicet quod omnis operatio naturae est modo finitissimo, ordinatissimo, brevissimo et optimo quo ei possibile est'", Crombie, *Grosseteste*, p. 86, translating Grosseteste, *De iride seu de iride et speculo, Werke*, p. 75.
66 Crombie, *Grosseteste*, p. 122, Cf. Baur, *Philosophie Grosseteste*, pp. 114-115.
67 Baur, *Philosophie Grosseteste*, p. 128.
68 *Ibid.*, pp. 113-119 for a further explanation and diagrams referring to the burning glass and the angles of refraction.
69 Cf. Anselm of Canterbury, *Proslogion* in *A Scholastic Miscellany: Anselm to Ockham*, Library of Christian Classics, Vol. X (London: SCM, 1956) XVI, 84.
70 Cf. St. Augustine *De Trinitate* in *Augustine: Later Works*, Library of Christian Classics, Vol. VIII (London: SCM, 1955) XII, 93, where, according to Augustine, created truth is known insofar as the knower is provided with the light of the *ratio aeterna* ; Anselm, *Proslogion*, XVI, 84.
71 Baur, *Philosophie Grosseteste*, p. 207.
72 Crombie, *Grosseteste*, p. 10. Cf. Nebelsick, *Circles of God*, pp. 126 ff.
73 Crombie, *Grosseteste*, p. 14.
74 *Ibid.*, p. 10.
75 Robert Adamson, *Roger Bacon The Philosophy of Science in the Middle Ages* (London: Simpkin, 1876), p. 8.
76 Crombie, *Grosseteste*, p. 140. Aristotle's *Prior Analytics* is devoted to discussing the theory and use of the syllogism while his *Posterior Analytics* concerns the acquisition, demonstration, expansion and systematization of knowledge, *i.e.*, epistemology. Cf. Roger Bacon, *Opus Majus*, tr. Robert Burke, 2 vols. (Philadelphia: University of Pennsylvania, 1928), I, Part Four, esp Ch. II and III, pp. 117-127; Ch. I and II, pp. 139-147; and II, Part Six, esp. Ch. I-III, pp. 583-589.
77 Cf. Nebelsick, *Circles of God*, pp. 131 ff. and Bacon, *Opus Majus* I, Part

4, Ch. XVI, pp. 201, 394; I, Part 1, Ch. III, p. 10. Bacon cited Aristotle throughout his *Opus Majus* and records that he was appalled both by objections to the philosophy and metaphysics of Aristotle as interpreted by Avicenna and Averroës and by the fact that those who used them had been excommunicated for long periods. *Ibid.*, p. 22.

78 *Ibid.*, I, Part IV, Dist. 4, p. 405.

79 Adamson, *Bacon*, p. 9. At the same time, Bacon's statement that Grosseteste neglected all the books of Aristotle is obviously contradicted by the evidence. Roger Bacon, *Compendium studii philosophiae, Opera*, Rolls Series XV (London: Longman, 1859), cap. VIII, p. 469.

80 According to Adamson, he experimented with Friar Thomas Bungay and with him acquired a reputation as a dealer in "the Black Arts". Adamson, *Bacon*, p. 16. Cf. Bacon, *Opus Majus*, II, Ch. XII, pp. 626 f. For a discussion of Bacon's imprisonment and release, cf. Adamson, *Bacon*, pp. 16 ff.; H. Stanley Redgrove, *Roger Bacon, the Father of Experimental Science and Medieval Occultism* (London: Rider, 1920), pp. 19-29; J. H. Bridges, *Roger Bacon* (London: Williams and Norgate, 1914), pp. 31-35; S. C. Easton, *Roger Bacon and his Search for Universal Science* (Oxford: Blackwell, 1952), pp. 186-205. For an account of Bacon's "natural magic", cf. J. S. Brewer, ed., "Life of Roger Bacon (From Wood's Antiquitates Univ. Oxon.)", *Chronicles and Memorials of Great Britain*, Rolls Series, No. XV (London: Longman, 1859), pp. xci-xcvi and *ibid.*, Appendix I, "Epistola Fratris Rogerii Baconis de Secretis Operibus Artis et Naturae, et de Nullitate Magiae", pp. 523-551.

81 F. Sherwood Taylor, *The Alchemists* (New York: Collier, 1962), esp. Ch. 8-11, pp. 82-125. Taylor indicates that Chester's translation was followed by half a dozen translations of other Arab alchemistic texts before the end of the century, *ibid.*, pp. 83 f. As an explanation of alchemistic conceptualities, Taylor refers to Raymond Lull (c.1235 - c.1316), one of the foremost of medieval alchemists who explained that the thing which God created was what he called "argent vive" (*argentum vivum*, quicksilver, mercury), and that this original matter gave rise to all other things. The finest part formed the bodies of the angels, a less fine part the heavenly spheres, stars and planets, and the coarsest formed the terrestrial bodies. In the terrestrial bodies part of this "argent vive" became the four elements—earth, water, air, and fire—but a part remained as a fifth element, the *quintessence*. Thus, in every body there was some stuff akin to the heavenly bodies, and it was through this material that the heavenly bodies could bring about the changes of generation and corruption. The activity of the body abode in the quintessence and alchemy was a process dealing with this fifth element and attempting to multiply its activity. *Ibid.*, pp. 93-95. Cf. Nebelsick, *Circles of God*, pp. 162 ff.

82 Newton is known to have kept his fires burning at Cambridge and made close observation of the effects of heat upon metals and

chemicals. The quantification and careful recording of chemical combinations and reactions in the short-term led to useful medicines and in the long-term to experimental chemistry. The recombination of atomic structures into new elements may well now be within the possibility of modern physics.

83 Cf. Nebelsick, *Circles of God*, pp. 168 ff.
84 Redgrove, *Bacon*, pp. 15 f.
85 Adamson, *Bacon*, p. 15. By contrast Bacon's assessment of his teachers, Adam Marsh and Robert Grosseteste, at Oxford can be judged from the fact that, according to Adamson, he calls them "the wisest of moderns" and commended them for being "perfect in both science and philosophy". *Ibid.*, p. 9.
86 Roger Bacon, *Opus Tertium, Opera*, ed. J. S. Brewer, Rolls Series XV (London: Longman, 1859), cap. xiii, p. 46. Cf. C. A. Burland, *The Arts of the Alchemists* (New York: Macmillan, 1967), p 2. Cf. also, S. P. Thompson, "Petrus Peregrinus de Maricourt and his 'Epistola de Magnete'", *Proceedings of the British Academy* (London, 1906), pp. 377-408.
87 Bacon, *Opus Majus*, II, Part VI, Ch. I, p. 583.
88 *Ibid.*, I, Part IV, lst. Dist., Ch. 3, p. 126.
89 *Ibid.*, I, Ch. 1, p. 116.
90 *Ibid.*, p. 117. Cf. Aristotle, *Metaphysics* I, LCL (London: Heinemann, 1933) IV. i, 1026a.
91 Bacon, *Opus Majus*, I, 1st Dist., Ch. II, p. 120; cf. *Ibid.*, pp. 118-120.
92 *Ibid.*, Part IV, 2nd Dist., Ch. I, p. 128-130.
93 *Ibid.*, p. 128.
94 *Ibid.*, Ch. II, p. 131.
95 *Ibid.*, p. 132.
96 *Ibid.*, p. 133.
97 *Ibid.*, Ch. III, pp. 136 f.
98 *Ibid.*, p. 138.
99 *Ibid.*, 3rd Dist., Ch. III—4th Dist., Ch. I-IV, pp. 146-158.
100 *Ibid.*, 4th Dist., Ch. VI, pp. 160 f.
101 *Ibid.*, Ch. VII, p. 163.
102 *Ibid.*, II, Part V, 4th Dist., Ch. III, p. 446. Cf. *Ibid.*, pp. 419–452. Adamson points out that, although Bacon devotes the fifth part of his *Opus Majus* to optics reporting on vision in a right line, laws of refraction and reflection and of the construction of mirrors and burning glasses, all of which he explicates mathematically, he does not move beyond Alhazen. Adamson, *Bacon*, p. 28.
103 Bacon, *Opus Majus*, II, Part V, 3rd Dist., Ch. III, p. 441.
104 Plato, *Republic*, Bk. VI. 510.
105 *Ibid.*, Bk. VI. 511.
106 Taylor, *Platonism*, p. 39.
107 Plato, *Republic*, Bk. VI. 511.
108 *Ibid.*, Bk. VI. 510.
109 Crombie, *Grosseteste*, pp. 155 f.

110 Bacon, *Opus Majus*, II, Part VI, Ch. II, p. 587.
111 *Ibid.*, p. 588.
112 The explanation, as Crombie points out, is given by Aristotle, *Meterologica*, LCL (London: Heinemann, 1952), III. v, 375b f. Bacon confirmed his deductions with astrolabic measurements showing that the sun was always in a straight line with the centre of the bow and the observer's eye. Crombie, *Grosseteste*, p. 157. Hence, the bow would be seen at different altitudes at different latitudes on earth and never when the sun was higher than 42° in the sky. Bacon *Opus Majus*, II, Part VI, Ch. V, pp. 593-596.
113 *Ibid.*, Ch. VII, p. 599.
114 *Ibid.*, pp. 602 f.
115 *Ibid.*
116 *Ibid.*, Ch. VIII, p. 605.
117 *Ibid.*, Ch. XII, p. 611.
118 *Ibid.*, p. 615.
119 Redgrove, *Bacon*, p. 26.
120 "Hermes Mercuris, pater philosophorum". Roger Bacon, *Opus Minus, Opera*, Rolls Series XV (London: Longman, 1859), p. 313. For Thorndike's discussion of Bacon and Hermetic lore, cf. Lynn Thorndike, *A History of Magic and Experimental Science*, 8 vols. (London: Macmillan and New York: Columbia University, 1923-1958), II, 219.
121 Albert the Great, *Speculum Astronomiae, Opera*, Vol. X, ed. Borgnet (Paris: Vivès, 1891), cap. XI, p. 641. Cf. Frances Yates, *Giordano Bruno and the Hermetic Tradition* (London: Routledge, 1964), p. 48. Thorndike, *History of Magic*, II, 220.
122 Burland, *Arts of the Alchemists*, p. 2.
123 The anthology was published by Macé Bonhomme at Lyons in 1557.
124 According to the frontispiece the work was "Composed by the thrice-famous and learned Fryer Roger Bacon". Roger Bacon, *The Mirror of Alchimy* (London: Olive, 1597). For discussions of the authorship of *The Mirror of Alchimy*, cf. Redgrove, *Bacon*, pp. 37 f., and Thorndike, *History of Magic*, V, 537.
125 Bacon, *Mirror of Alchimy*, p. 1.
126 Rather than being the elements of mercury and sulphur as we know them, for the alchemists "mercury" was an ethereal and very pure fluid which controlled everything from nutrition, sense, motion, power and colours to the retardation of age. "Sulphur" was a sweet, oil-like and viscous balsam which conserved the natural heat of vegetation and, among other things, was the instrument of its increase and transmutation. Taylor, *Alchemists*, pp. 156 f., Cf. Nebelsick, *Circles of God*, pp. 168 ff.
127 Bacon, *Mirror of Alchimy*, p. 7.
128 Roger Bacon, *Communium Naturalium, Opera*, 16 vol. in 12, ed. Robert Steel (Oxford: Clarendon, 1905-1940), Fas. II, Pars. I, Dist. ii, cap. 6, p. 90.

129 *Ibid.*
130 Adamson, *Bacon*, p. 28.
131 According to Adamson, Bacon was convinced that anyone attempting to do science and did not work with mathematics, optics and speculative alchemy laboured in vain. Adamson, *Bacon*, p. 15.
132 Easton, *Bacon*, p. 181.
133 *Ibid.*, pp. 181 f.
134 *Ibid.*, pp. 182 f.
135 Bacon, *Opus Majus*, II, Part VI, Ch. XII, pp. 629 f.
136 Cf. H. W. L. Hime, *Gunpowder and Ammunition* (London: Longmans, 1904) and Hime's contribution to the essays on the subject edited by A. G. Little (Oxford, 1914). Cf. also Redgrove, *Bacon*, p. 61 where he claims that Bacon actually invented gunpowder in contradistinction to the Chinese "incendiary" mixtures.
137 Easton, *Bacon*, p. 179.
138 *Ibid.*, p. 74. Cf. Bacon, *Opus Majus*, I, Part II, Ch. I-XIX, pp. 36-74.
139 For a discussion of both Bacon's and Arab Hermeticism as well as a discussion of Hermeticism in general, cf. Nebelsick, *Circles of God*, pp. 162 ff.,168 ff.,173 ff.
140 Easton, *Bacon*, p. 230.
141 Cf. Adamson, *Bacon*, pp. 29 ff.
142 *Ibid.*, p. 6; cf. *Ibid.*, pp. 29 ff.
143 T. F. Torrance, "Intuitive and Abstractive Knowledge from Duns Scotus to John Calvin", *Acta of the Duns Scotus Congress, 1966*, ed. P. Balicv (1968), IV, 291.
144 Cf. Crombie, *Grosseteste*, p. 167.
145 Crombie refers to John Duns Scotus, *Quaestiones in Lib. I Sententiarum* (London,1639), Dist. iii, Q. 4, p. 479 Cf. Crombie, *Grosseteste*, p. 168.
146 John Duns Scotus, *Metaph.* VIII, cited by Torrance, "Duns Scotus", p 292.
147 Aristotle, *Posterior Analytics*, LCL, tr., Hugh Tredennick (London: Heinemann, 1960), 71a8, cf., 81b2, 88a4.
148 John Duns Scotus, *Priorum Anal. Quaest* ii, Q. 8, text in Crombie, *Grosseteste*, pp. 168 f., n. 6. Cf. Aristotle, *Prior Analytics*, II. xxiii, 68b.
149 John Duns Scotus, *Quaest. Sent.* i. iii. Q. 4, pp. 482-483, cited in English, Crombie, *Grosseteste*, pp. 169-171
150 Crombie, *Grosseteste*, pp. 169 ff.
151 *Ibid.*, p. 170.
152 *Ibid.* Cf. Torrance, "Duns Scotus", pp. 292 f.
153 John Duns Scotus, *Quaest. Sent.* i. iii. Q. 4, pp. 482-438.cited in English, Crombie, *Grosseteste*, p. 170.
154 Cf. Crombie, *Grosseteste*, p. 167, n. 7. Cf. above, p. for "the principle of economy".
155 Böhm, *Philoponos*, especially Ch. VII, "Wirkung des Johannes Philoponos auf die Nachwelt" ("The Influence of Johannes Philoponos on the World after Him"), pp. 337-387.
156 Saint Basil, *Exegetic Homilies*, *The Fathers of the Church*, Vol. 46 (Wash-

ington D.C., Catholic University, 1963), pp. 6 f.
157 Saint Augustine, *The Confessions of St. Augustine* (New York: Sheed, 1943), XII. 17, pp. 302 f.
158 *Ibid.*, XIII. 15, p. 331.
159 Sambursky, *Physical World of Late Antiquity*, pp. 160-162.
160 Venerable Bede, περὶ διδάξεων sive *Elementorum Philosophiae*, (*Dubia et Spuria*), *Opera Omnia*, Tomus Primus, Patrologia Latina, Vol. 90 (Paris: Migne, 1862), Liber Primus, cols. 1132-1137. For a full discussion, cf. Nebelsick, *Circles of God*, "Science Encounters the Christian Faith", pp. 88-148. As early as the sixth century B.C., the pre-Socratic philosopher Anaxagoras (c.500 - 428 B.C.) who was too early to be caught up in the Platonic- Aristotelian ideas of dividing the heavenly incorruptible sphere from the corruptible earthly one, based his evidence on the composition of a meteorite that the heavenly bodies may have been composed of stone-like material.
161 Whitehead, *Science and the Modern World*, p. 13. n
162 Cf. Nebelsick, *Circles of God*, pp. 262 ff.
163 Crombie, *Grosseteste*, p. 175.
164 *Ibid.*, p. 172. Crombie's statement that Ockham strongly attacked Duns Scotus and attempted to return to Aristotle in this context is difficult to understand.
165 Crombie, *Augustine to Galileo*, II, 48.
166 *Ibid.*, p. 318.
167 Torrance may well be right in asking at this point, "Is Crombie not interpretating Ockham in a too Humean and modern light?" Thomas F. Torrance, *Theological Science* (London: Oxford University, 1969), p. 64, n. 3.
168 William of Ockham, *Lib. Sent.*, Prol. Q. 2, M cited by Crombie, *Grosseteste*, p. 173. William of Ockham *Super Libros Quatuor Sententiarum*, bk. l, dist. 45 ques. I,D, cited by Crombie, *Augustine to Galileo*, II, 45 f.
169 Ockham, *Lib. Sent.*, i. xlv, Q. l,D. Text and translation by Crombie, *Grosseteste*, p. 173.
170 "Ut frequenter, non potest evidenter cognosci aliqua res singularis contingens sine multis apprehensionibus singularium." William of Ockham, *Summa Logicae, Opera Philosophica*, Vol. 1, ed. P. Boehner (St. Bonaventure, N.Y.: St. Bonaventure University, 1974), iii. ii. 10, pp. 523 f. Cf. Crombie, *Grosseteste*, p. 173.
171 Crombie, *Grosseteste*, p. 174. Emphasis ours.
172 Ockham, *Lib. Sent.*, Prol. Q. 2,G. Translated by Crombie, *Grosseteste*, p. 174.
173 *Ibid.* Text and translated by Crombie. Italics added.
174 William of Ockham, *Quodlibeta*, iv, cited by Crombie, *Grosseteste*, p. 174. Crombie also refers to William of Ockham, *Summula in Libros Physicorum*, ii. 6. Crombie, *Grosseteste*, p. 174, fn. 2. Cf. also Crombie, *Augustine to Galileo*, II, 45 f.
175 Ockham, *Lib. Sent.* ii. Q. 3,G cited by Crombie, *Grosseteste*, p. 174.

176 Crombie, *Grosseteste*, p. 175.
177 *Ibid.*
178 L. O. Kattsoff, "Ptolemy and Scientific Method", *Isis*, xxxviii (1947), pp. 18-22.
179 Crombie, *Grosseteste*, pp. 85 f., n. 4.
180 "Quia pluralitas non es ponenda sine necessitate. William of Ockham, *Quodlibeta Septem, Opera Theologica*, Vol. IX (St. Bonaventure: St. Bonaventure University, 1980), Quod. V, Q. 5, p. 495. Crombie says the well-known "Ockham's razor" phrase, "Entia non sunt multiplicanda praeter necessitatem" ("Entities are not to be multiplied beyond necessity"), was introduced in the seventeenth century by John Ponce de Cork, a follower of Duns Scotus. Crombie, *Augustine to Galileo*, II, 45 and Crombie, *Grosseteste*, p. 167, n. 7.
181 Einstein dispensed with his "cosmological constant", for instance, on the basis of economy. Cf. Einstein, *The Meaning of Relativity*, (London: Chapman and Hall, 1980), pp. 120 f.
182 Ockham, *Sentenzenkomm.*, qu. 22, cited by Böhm, *Philoponos*, p. 358. Both Aristotle and Roger Bacon had denied the existence of a vacuum or "void" calling it "an obstruction". Aristotle, *Physics* II, IV. vi. 213b-216b; Bacon *Opus Majus*, II, Part V, Dis. IX, Ch. I, p. 485. Interestingly enough Einstein in his theory of relativity agrees with Aristotle with regard to "the void" in relationship to space. For Einstein, as for Aristotle, there is no differentiation between an object and the space it occupies. Einstein, *Meaning of Relativity*, p. 3.
183 As indicated above, Philoponos was the first to break with Aristotle's concept of motion.
184 Ockham, *Sentenzenkomm.* II, qu. 18 and 26 cited by Böhm, *Philoponos*, p. 359. Cf. Crombie, *Grosseteste*, p. 176 and *Augustine to Galileo*, II, 75-79.
185 William of Ockham, *Tractatus de Successivis*, ed. P. Boehner (St. Bonaventura: Franciscan Institute, 1944), p. 45.
186 Crombie, *Augustine to Galileo*, II, 46.
187 *Ibid.*, pp. 46 f.
188 *Ibid.*, p. 46.
189 Torrance, *Theological Science*, p. 63.
190 William of Ockham, *Comm. in Sent.* Prol. qu. 1 QQ: ii q. 150; iii q. 8d and *Quodlibeta* i, xiii-xv referred to by Torrance, *Theological Science*, p. 63. It should be pointed out that while in his *De Aeternitate Mundi* Thomas insists as over against Averroës on the doctrine *creatio ex nihilo* which he translates *creation after nothing*, (Aquinas, *Eternity of the World*, p. 22), he continued to insist that God's rational nature so interpenetrated the world that God not only caused the world (*Ibid.*, p. 54) but, since nature was shot through with ultimate causes, God and his eternal pattern could be read off the face of creation. St. Thomas Aquinas, *The Summa Theologica*, 3 vols., tr. Fathers of the English Dominican Province (New York: Benziger, 1948), I, q. 8, a. 1-3; q. 14, a. 13; q. 19, a. 4-12; q. 44, a. 1; q. 57, a. 3.

191 William of Ockham, "An Excerpt from Eight Questions on the Power of the Pope", *A Scholastic Miscellany: Anselm to Ockham*. The Library of Christian Classics, Vol. X (Philadelphia: Westminster, 1956), p. 442.
192 Torrance, *Theological Science*, p. 63.
193 William of Ockham, *Comm. in Sent.* Prol. qu. 1 QQ: ii q. 150; iii q. 8d and *Quodlibeta* i, xiii-xv referred to by Torrance, *Theological Science*, p. 63. It should be pointed out that while in his *De Aeternitate Mundi* Thomas insists as over against Averroës on the doctrine *creatio ex nihilo* which as we have seen he translates *creation after nothing*, he continued to insist that God's rational nature so interpenetrated the world that God not only caused the world (*Ibid.*, p. 54) but, since nature was shot through with ultimate causes, God and his eternal pattern could be read off the face of creation. St. Thomas Aquinas, *The Summa Theologica*, 3 vols., tr. Fathers of the English Dominican Provence (New York: Benziger, 1948), I, q. 8, a. 1-3; q. 14, a. 13; q. 19, a. 4-12; q. 44, a. 1; q. 57, a. 3.
194 Jean Buridan, *Quaestiones in octo libros Politicorum Aristotelis* (Biblioteca Apostolica Vaticana, 1509), folio cxxi, quaes. 12. Cf. Pierre Duhem, *Études sur Léonard de Vinci*, 3 vols. (Paris: Hermann, 1906, 1909, 1913) III, p. 42; Böhm, *Philoponos*, pp. 349 f.
195 Sambursky, *Physical World of Late Antiquity*, p. 71. Cf. above, pp.
196 Cf. A. R. Hall, *From Galileo to Newton 1630-1720* (London: Collins, 1963), esp. pp. 36 ff., for an explanation of the inadequacy of empiricism. Hall's terms "obstruction" and "idealism" in this regard may be unfortunate, however, as to how science proceeds.
197 Whitehead, *Science and the Modern World*, p. 13.
198 Plato, *Timaeus*, 34.
199 Cf. Paul Tillich, *The Construction of the History of Religion in Schelling's Positive Philosophy* (Lewisburg, PA: Bucknell University Press, 1974). Cf. F.W.J. von Schelling, *System des transcendentalen Idealismus* (1800) Cf. Diogenes Allen, *Philosophy for Understanding Theology* (Atlanta: John Knox, 1985) where he points to the necessity of the necessary differentiation between creator and creation for Christian theology. Cf. A.N. Whitehead, *Process and Reality* (New York: Social Science Book Store, 1941), pp. 521 ff. Cf. Nebelsick, *Theology and Science in Mutual Modification* (New York: Oxford University Press, 1981), pp. 45-62.
200 Plato, *Timaeus*, 30. Thus Plato reflects the Babylonian myth of creation being formed by the god Marduk, the god of order who first defeated Tiamat, the goddess of chaos and then formed the world out of the primordial chaos itself.
201 *Ibid.*, III, 36, p. 619. Nicholas of Cusa's modification of the statement is, "A circle whose center is everywhere and whose circumference is nowhere." (Nikolaus von Kues, *De docta ignorantia*, ed., Paul Wilpert (Hamburg: Verlag Felix Meiner, 1970-77), p. 95.
202 Cf. Torrance, *Theological Science*, p. 59 who refers to M. B. Foster, "The Christian Doctrine of Creation and the Rise of Modern Science", *Mind* xlii (1934), pp. 446 ff.; "Christian Theology and Mod-

ern Science of Nature",Parts I and II, *Mind* xliv (1934), pp. 439 ff., and xlv (1936), pp. 1 ff.; *The Christian News Letter* 299 (Nov. 1977) and also M. B. Foster, *Mystery and Philosophy* 1957, pp. 87 ff.

Chapter 2

THE RENAISSANCE MIND

The Roots of the Renaissance
The Renaissance was a strange and esoteric mixture.[1] It arose when under the influence of Arab culture European intellectual development was set afire through the introduction of Greek philosophy. There was a great revival of classical thought. This was a uniquely foreign if not esoteric revolution within intellectual history. Eventually it was just this amalgamation of exotic ideas which was to prepare for both the Reformation and the rise of modern science. Its various elements were to inspire great changes. Aristotle's works: *On the Heavens* and *Physics* re-awoke interest in nature. The *Metaphysics* and the *Prior Analytics* introduced the deductive, logical side of Aristotle. Euclidean geometry and Ptolemy's (c. 90 - c.168) *He Mathematike Syntaxis* opened the world to order and measurement. Eventually Aristotle's *Posterior Analytics* set out the pattern of inductive reasoning. At the same time, the esoteric Pseudo-Dionysian writings kept alive and encouraged both the sense of mystery about the universe and the practice of the occult that had been a characteristic of thought influenced by neo-Platonic philosophy from Augustine onwards.

Basic to this revival was a belief in the authenticity of the *ancient* as over against the *contemporary*. This inspired the search for and the gathering of Plato's *Dialogues* and the *Corpus Hermeticum* by Cosimo di Medici (1389-1464) in Florence as well as their translation by Marsilio Ficino (1433-99), director of the Florentine Academy. Ficino translated the Hermetic writings in 1463 and Plato's *Dialogues* in the years 1467-69. So popular was the *Corpus Hermeticum* that it went through sixteen editions by the end of the sixteenth century. Esoteric though they were, or perhaps just because they were esoteric, they facilitated the reassertion of Platonism over the dogmatic

Aristotelianism of Thomist thought.

The Hermetic emphasis on the ephemeral and the mystical, that fed the almost frantic belief in the occult, was complemented by the other side of Aristotle, his *astrology*. It taught that destiny was controlled by the stars. Astrology was accompanied by her cultic sister, *alchemy*, who promised wealth by turning quick-silver into gold and encouraged the search for the essence and elixir in all things, the philosopher's stone. If found and distilled, the philosopher's stone would insure the health of the body and the regeneration of the soul.

The quest for ancient truth lead to a re-emphasis and re-examination of the ancient documents not only of philosophy and science but of the Christian faith as well. The Scriptures of the Old and New Testaments were re-emphasized and re-interpreted. The discovery of gross discrepancies between the teachings of the ancient church as found in the documents of Scripture and the doctrine and practice of the then current church was the impetus that led eventually to the Protestant Reformation. Accordingly, the mind-forming and culture-determining influences of the Renaissance that set the parameters within which the Reformation arose and modern science began, were the teachings of Aristotle, Plato, neo-Platonism, Hermeticism and the Christian faith as based upon Scripture.

The development of theology in the West would tend, by and large, to bear out the testimony of John Marsh, "It would seem to be as characteristic of the Reformed theologian to follow Plato as for the Catholic to follow Aristotle."[2] Whether or not Thorlief Boman is entirely correct in stating that "Platonism and Christianity are related essentially" and that "they support and join values",[3] it is certainly true that the residue of Platonism in the West, along with that which was rediscovered in the Renaissance in both the Platonic writings and in the form of Hermeticism, served as an impetus for the Reformation. These same sources helped to instigate the late Renaissance movement against the Aristotelianism that had been adopted and propagated by the Thomist theology of the church of Rome.

It would be wrong, however, to differentiate Plato and Aristotle too sharply. Both inherited the Pythagorean conceptuality of the heavens that moved in perfect circularity and were thought to be divine, eternal, and immutable. Both described the action of the god-like planets as animated and as being analogous to animals and plants. Both differentiated the supra-lunar, immutable, heavenly, divine, and rational realm, the place of perfect circular motion from the sub-lunar, material sphere, the place of change, rectilinear motion, and irrationality. Both believed in the occult capacities of the stars.[4] Consequently neither valued the material world as a proper object of philosophical investigation. Plato was willing to consider material as only a manifestation of the *ideal form* that lay behind it. Aristotle conceived of the material as an *accident* of the *real substance*. Real substance was a result of the first and final causes that alone gave the material its essence. He criticized the Ionian natural philosophers, the *physikoi*, for not attempting to penetrate behind material phenomena to their causes.[5]

The very success of Aristotle in the classification of plants and animals persuaded him to view the whole of nature *in terms of organismic wholes*. He considered "the formal nature" to be of greater fundamental importance than "the material nature".[6] Consequently, he concentrated on wholes rather than on the constituent parts of the whole. This attitude mitigated against his ability to understand the cosmos in physical terms.[7] "Final causes" rather than immediate ones determined nature. All motion "has the end for its purpose".[8] In fact, God and nature do not create anything that does not fulfill a purpose.[9] In the end, then, the whole of reality was understood *teleologically*. It was "end-oriented" and since the nature of the end was at one and the same time both speculative and determinant, the whole of Aristotelian science was forced into a box of its own making.[10] Thus, rather than following the Ionian *physikoi*, who had begun to try to understand nature as nature, and had made an attempt at considering the whole in the light of the particular, Aristotle's effort forced nature to conform to presupposed first and final causes. Utilizing his own syllogistic method, he then deduced a description of nature that conformed to his own principles.

Thus, Aristotelian thought, had both positive and negative results in the West as far as the development of science was concerned. Both Aristotle's metaphysics and his physics served to revive the interest of the western mind in the physical world. His emphasis on the primacy of the deductive method, however, allowed science to develop only so far and no further. His focus on the general rather than the particular, and his insistence that phenomena were manifestations of first and final causes, all mitigated against the development of the empirical habit of mind whereby causes and relationships are arrived at inductively from observation.

In this way Aristotelian thought was "science-preventing". However, in that Aristotle's thought had the effect, at least, of stimulating interest in nature, the renowned mathematician E. T. Whittaker (1873-1956) is too harsh in his judgment that Aristotle's physics was "worthless from beginning to end".[11] It is true that Aristotelian physics was, if not worthless, certainly misleading. In fact, by providing wrong answers to legitimate questions, it tended to cut off inquiry that could have led to a valid method of investigation into nature *qua* nature. Yet the axiomatic method that was developed on the basis of Aristotle's *Posterior Analytics* manifestly inspired the movement toward science. His *hypothetical-deductive* method, which involved the formulations of hypotheses and then drawing out their implications by the application of deductive logic became manifest, for instance, in both Euclid's (fl. c. 300 B.C.) *Elements of Geometry* and in the writings of Ptolemy, who followed Euclid's method in the development of his cosmology. This method is best illustrated by Euclid's plane geometry. It set out a minimum number of basic postulates and then used Aristotle's deductive logic to derive the theorems and figures that were supposed to define the shape of space.[12] Ptolemy's *Almagest*, is a superb example of the application of the Aristotelian hypothetical-deductive method at its best. Being convinced of Euclid's method, Ptolemy set out his axioms, drew his theorems from them and worked out his astronomy accordingly.

However, whereas in the first instance the theorems are used to represent the "objects" of reality (in Euclid triangles, squares, circles, etc. and in Ptolemy to the positions and motions of the heavenly bodies), a subtle *epistemological inversion* took place as the systems became standardized. Rather than the theorems being used to explain the "objects", the objects were defined in terms of the theorems. This gave an impression of axiomatic completeness. The all-encompassing systems of Aristotle, Euclid and Ptolemy necessarily excluded those aspects of reality that failed to fit within their parameters. Thus, like all axiomatic arrangements, Aristotle's physics and metaphysics, Euclid's geometry and Ptolemy's geocentric cosmology performed a dual but dangerous function. They began as attempts to explain and give sense to reality but, as they later evolved, they excluded all "ideas" or "intuitions" which fell outside their compass and relegated these to "error", "irreality" or to "illusion".

Thus, when Euclid and Ptolemy, along with Aristotle's physics were made known in the West, the clarity, relative simplicity, logical consistency, and the completeness of their systems so opened the western mind to the understanding of nature that a revival of interest in nature as well as a thirst for ancient learning, in general, took place. However, in that the postulates of the systems were adequate only to the limited and closed understanding of reality predesignated by them, they both enabled the late medieval mind to understand the world and defined and limited the world in their own terms. As long as space was considered to conform to the formulae of Euclid, and the heavenly motions were considered to form perfect circles around the earth as explained by Ptolemy, the systems were both descriptive and productive. However, in those instances where the understanding of nature began to outrun the systems or run counter to them, they not only proved to be too simplistic to comprehend the complexity of reality but, by continuing to limit vision to their own parameters, they served to prevent investigation into nature beyond what we now judge to be a rather rudimentary level.

These systems, then, like all systems of thought were only as adequate as were the fundamental principles (axioms) on which they were based and the logic by which the axioms were connected and worked out. When their descriptions proved inadequate or just plain wrong and their logic was found to be too limiting, they, like all outmoded systems, not only became suspect, they became counter-productive. They then needed to be broken, supplemented or replaced with suppositions and logic that advanced the understanding of reality while allowing and encouraging thought opened to new disclosures of reality.

Because the compass of Aristotle's thought was much wider than were the mathematical arrangements of Euclid or Ptolemy, Aristotelianism was not a completely closed system as were Euclidean geometry and Ptolemaic cosmology. Whereas Euclid and Ptolemy were satisfied for the most part, to deal with descriptions of the "phenomena of nature" — Euclid with the description of space and Ptolemy with the description of the planetary system — Aristotle's effort was largely an attempt to deal with the metaphysical causes behind nature. Like the pre-Socratic Ionians, Aristotle was primarily concerned with the causes behind motion and change. In spite of the fact that Aristotle could criticise Plato for making the movement of the heavens dependent upon a cosmological spirit, he himself remained as convinced as was his master that the heavens themselves were divine and that the "godly stars" arbitrarily intervened in both worldly and human affairs.[13]

This mystical, *astrological* interpretation of reality (that had its roots in the early Pythagoreans, who influenced Plato) came into the West in an especially powerful form in the neo-Platonism of Plotinus (c.205 - c.270). In Plotinus we see the roots of the neo-Platonic system that were an amalgamation of Platonic ideas with compatible elements of Aristotelian metaphysics. The system was adapted and propagated by the theology of Augustine and in that form it was responsible for keeping Platonism alive in the West throughout the Middle Ages. When this residue of Platonism was re-enforced by the mystical hierarchies of a revived neo-Platonism, it began to undermine dogmatic Aristotelianism whenever the two encountered one another. As we have seen, Platonic ideas, along with Christian concepts, inspired Philoponos (c.490 - c.566) at Alexandria to challenge Aristotelian cosmology as early as the sixth century. So too, these same factors, were involved in the questioning and the rejection of much of Aristotle by both Robert Grosseteste and Roger Bacon in the late twelfth and early thirteenth centuries.

In the late Middle Ages, the mystical neo-Platonism of Augustine was complemented by the even more esoteric, liturgical and mystical conceptualities of the Pseudo-Dionysian writings. The sixth-century writings based primarily on the neo-Platonism, as represented by Proclus' (c.410 - 485) *Elements of Theology*, served to "baptize" the neo-Platonic mystical understanding of the universe into Christianity. When in the thirteenth century Thomas Aquinas incorporated the teachings of Pseudo-Dionysius *The Celestial Hierarchies* into his *Summa Theologica*, and when Dante Alighieri (1265-1321) used them as the basis of his dramatic works, *The Banquet* and *The Divine Comedy*, the neo-Platonic occult system became the very foundation of late medieval and early Renaissance piety.

The writings of Pseudo-Dionysius were authoritative in part, no doubt, because they were introduced to the West under the guise of apostolic tradition. The pseudonymous author wrote under the name of Dionysius the Areopagite, thereby identifying himself with the Dionysius of Acts 17:34, one of the few converts to follow the Apostle Paul after he had delivered the sermon at the Areopagus in Athens. Although the authenticity of the Dionysian writings was first questioned when they were introduced at the beginning of the sixth century and again occasionally in the Middle Ages, it wasn't until the sixteenth century, concurrent with the rise of Protestantism, that considerable controversy challenged the authenticity of their authorship. It was not until 1895 that indisputable evidence was put forth regarding the relatively late date of the Pseudo-Dionysian corpus.[14] Thus it is difficult to fault the late medieval writers for believing in their authenticity.

In that there are distinct parallels between the Pseudo-Dionysian understanding of the world, Augustinian neo-Platonism, and the astrological elements of Aristotle, it isn't surprising that Thomas Aquinas could accept the Pseudo-Dionysian "Christian apology". Although the nine orders of angels, each divided into groups of three within each order, with each group of three related to one of the three Persons of the Trinity, have little in common with biblical teaching, the arrangement fits hand in glove with the mystical elements of both Augustinianism and Aristotelianism that form the foundation of the medieval synthesis of Thomas Aquinas. Thus, like the even more exotic neo-Platonic Hermeticism on which we will touch below, the esoteric cultic system of Pseudo-Dionysius was introduced to the late Middle Ages with the best of references.

Even today neither the authorship nor the place of origin of the Pseudo-Dionysian writings has been firmly established. Since the earliest extant citation from them was made in the early sixth century, the present judgment is that it is highly improbable that they existed earlier. The record indicates that they were first referred to in Constantinople in 532 and were translated into Syriac shortly afterward. In the early sixth century John of Scythopolis (fl. 6th cent.) wrote a commentary on the Dionysian writings. The first record of their introduction to the West goes back to 754 when Pope Paul I (d. 767) sent a Greek copy to Pepin the Short (715-768), the Frankish king and the father of Charlemagne (742-814). A second copy was sent from the Byzantine emperor Michael II (d. 829) to Louis the Pious (778-840). This work was translated into Latin by Abbot Hilduin of St. Denis (d. c.842). Later, John Scotus Erigena (c.810 - c.877) made a more accurate translation. Commentaries were made by Maximus the Confessor (c.580-662) and John Colet (c.1467-1519).[15] Thus Raymond Klibansky is quite correct in judging that the Dionysian writings occupied an "eminent position" throughout the

Middle Ages.[16]

It is understandable, therefore, that Thomas Aquinas, whose plethora of citations from Pseudo-Dionysius are second in number only to his citations from Aristotle, could call upon the Pseudo-Dionysian "Christianized" neo-Platonism of Proclus as the basis for his medieval piety. The hierarchies of angels that bridge the gap between heaven and earth are the Pseudo-Dionysian personalization of the "transcendent intelligences" described by Proclus. These guide the souls, which are incarnate earthly bodies, from heaven to earth and back again to their final heavenly home where they live forever in the presence of God himself in whose being they participate.[17]

Thomas' mystical piety, written in the prosaic style of medieval scholastic treatises, was complemented and supplemented by Dante, who translated the Proclus-inspired Dionysian mystical world of angelic hierarchies into the prose and poetry of his *The Banquet* and *The Divine Comedy*. In so doing, Dante linked Aristotelian cosmology, Dionysian angelology and medieval theology into a mythical, mysterious and awe-inspiring infinite chain of hierarchically-arranged beings who mediated divinity from God above through orders of angelic beings to creatures below. Divine impulses initiated from heaven were imputed into souls on earth until, lifted by divinity, they returned home to the "Empyrean", the dwelling place of God and of the "spirits beatified".[18]

It is thus not surprising that Marsilio Ficino, the most renowned Platonist of the Renaissance, could find the Pseudo-Dionysian writings inspirational. In his *De Christiana Religione* (*On the Christian Religion*), Ficino portrays cosmic order in analogy to the Pseudo-Dionysian hierarchies. Like Thomas Aquinas, he has nine orders of angels that reach upward from the earthly "elements" to the "Empyrean". Also like Thomas Aquinas, he divides the orders into triads and relates each triad to one of the persons of the Trinity. Following Pseudo-Dionysius, Thomas and Dante, he delegates the responsibilities of the angels so that those above are in direct relationship to God while those below minister to people. The lower angels look after human affairs. As *guardian angels*, they care for those to whom they have been assigned.[19]

Thomas and Dante may form the bridge over which Dionysian mysticism moved throughout the Middle Ages. It was with Ficino, however, that Platonism, Hermeticism and other mystical versions of neo-Platonism were introduced into the world of the Renaissance. Ficino translated the *Corpus Hermeticum*, and writings of Plato, Plotinus, and Proclus. Ficino also combined these influences into the neo-Platonic Hermetic "new science" that helped to prepare the stage for the development of what we recognize as "modern science". In themselves these Platonic, neo-Platonic and even pseudo-Platonic influences were anything but scientific. However, when their impact arose during the Renaissance, their mystic, esoteric and erotic characteristics so influenced the Renaissance mind that they helped to break the back of the sterile Aristotelian rationalism. In so doing they encouraged the kind of thinking that was to lead to the development of modern science.

The Renaissance encyclopaedists such as Paracelsus (c.1490-1541), Jerome Cardan (1501-1576), Francesco Patrizzi (1529-1597), and Peter Severinus (1549-1602) were quick to appreciate this strange mixture of philosophy, astrology, and mystical occult piety. By relating and interweaving the exotic aspects of mystical Platonism with their own thought, they successfully incorporated it, stirred it into appropriate brews according to their own recipes, and served it to a populace that had been taught to hunger for the kind of miracles that myth promises.

Thus if the Achilles' heel of Aristotelianism was the straight-jacket of the deductive method based upon principles which *eo ipso* ruled out innovation, the weakness of Platonic and neo-Platonic thought was that it allowed for and even promoted flights of wildly speculative imagination. While such thinking helped to break dogmatic Aristotelianism by promoting speculation, it lacked a dependable descriptive concept of nature. It thereby encouraged a renewal of world mythicization. Thus, although the

Platonic legacy encouraged the Renaissance mind to use intuitive imagination in viewing reality, it is in itself highly suspect. To the rational, scientifically-oriented twentieth century mind (itself highly influenced not to say significantly narrowed by a reductionist vision propagated by popular modern science), the Platonic, neo-Platonic and neo-Pythagorean Hermetic writings seem incredulous and untrustworthy. However, they have played a significant role in the development of modern views on reality.

The mixing of the heavenly and the earthly, the divine and the human, the dynamic flow from heaven above to earth beneath and back again, would appear to be a fantasia rooted only in the imagination. The eternal harmony and unity of the heavenly sphere were contrasted with the disharmony, disunity, and non-reality of the material realm. As in both Plato and Aristotle, the rational was reflected in the perfect circularity of the divine, immutable, eternal, and celestial spheres. This resuscitation of ancient Pythagorean cultic concepts in Renaissance guise set the heavenly realm over against the irrationality of the sub-lunar, earthly sphere of rectilinear movement where generation, change, decay, and death reigned. In the sub-lunar sphere all things struggled to escape the bonds of the earth so as to rise again to the Empyrean, the heaven of heavens. There they would rest eternally and be at one with the One from which all in the first instance springs and to which all is destined to return. It followed as a matter of course that all things are presided over by souls, intelligences, or celestial beings. These governed each part of the world below directing its ways in preparation for its journey back to the *One*. There it would again be part of the *whole* out of which it originally emanated.

Although disconcerting and highly mythical, this cultic eccentricity identified the world and humankind with the gods and God and provoked the Renaissance mind to reach out beyond the *status quo* and to call it into question. This kind of belief structure identified souls with material objects and tied the fate of material things to the future of souls. It sought salvation not only for people but for the essence of all substances. Such a "faith" encouraged contemplation of that which was beyond the known but it revived an interest in nature as well, albeit a soul-filled nature. The untried, unthinkable, and hypothetical first became plausible and then possible. On the whole this kind of thinking filled the mind with unreality. In a few noteworthy instances, however, such esoteric contemplation beat at the doors of circumstance and helped to pry them open to the light that was to flood the western world during the eras of the Renaissance, the Reformation and the rise of modern science.

With a certain fear of reiterating the obvious, it is necessary to point out that Greek thought as it was transmitted to the Renaissance has its definitive *pre-history* in the age of Pericles (c.495 - 429 B.C.). That history, of course, was both manifested, developed, and propagated by Plato, on the one hand, and Aristotle, on the other. Since there is evidence to indicate that what we have of Plato and Aristotle is the result of editing and even invention, we are hardly in a position to argue that the extant writings of Plato actually represent the mind of Plato, or that the so-called Aristotelian corpus reflects Aristotle in pristine originality.[20]

However what we have is all we have. Whether or not the writings stem directly from Plato and Aristotle, they represent two emphases in Greek thought that, referring again to John Marsh, have affected western thought, somewhat in alternation with each other, from about the first century onward.[21] There was, as we mentioned, an older pre-Socratic Ionian philosophical tradition in Greece. Because it differentiated between the "spiritual" and the "material", it may well have pointed in a direction that was much more discriminatory and clear-headed than that of either Plato or Aristotle. However, Ionian philosophy lacked a "religious vehicle" to carry it through history. Consequently it dropped into the background and was by and large forgotten.

When Francis Bacon compared Plato and Aristotle to the pre-Socratics, he said that they "like planks of lighter and less solid material, floated on the waves of time, and were preserved".[22] It is certainly true that as Plato and Aristotle floated down the stream of western history, they pulled a good deal of that history including that of the Ionians in their wakes.²³ In the Renaissance it was Plato, who in various guises, reasserted himself over his pupil, Aristotle.

The Place of Plato
As we have indicated, even in Thomas Aquinas, Aristotelianism was complemented by, if not compromised by, the neo-Platonism of Augustine as well as that of Pseudo-Dionysius. A. E. Taylor goes so far as to say that Aristotle not only owes the best of his inspiration to the influences received from his personal contact with Plato, but that it is the Platonic strain in Aristotle that has had the greatest appeal to later ages.

> That "domination of the human mind by the authority of Aristotle" which popular imagination strangely exaggerates, is itself an episode in the fortunes of Platonism.²⁴

Taylor may have overstated the case, but there would seem little doubt that the Platonic understanding of the eternal as being the real and immutable as over against the material as being time-bound, changeable, and corruptible was as much a basis for Aristotelian thought as it was for that of Plato. Thus, an emphasis on the *Platonic side* of Aristotelianism could easily push the aspects of Aristotle's thought that were concentrated on material things into the background. Even in Aristotle, it was the eternal divine causes behind *the material* that were real. Thus *the physical* was in constant danger of falling back into the state of non-being.

For Plato, in particular, as indeed for the Pythagoreans from whom he inherited the basis for his thought and, as far as we know, for the Babylonians before them, philosophy, cosmology and religion were of a piece. It is no doubt because of this alluring but subtly devious amalgam that Plato, especially, was read with approval by Christian thinkers from the early Church Fathers onward. In the late Renaissance, Marsilio Ficino reflects the usual Christian apologetic attitude toward Plato by calling him a "religious philosopher". The early Church Fathers considered Plato a precursor of Christianity. It was because his writings were considered to be valuable in convincing people to adhere to the Christian faith that Plato was considered worthy of God's highest award, worthy even of eternal life.[25]

The two aspects of Plato's theologically-determined philosophy that were to affect the western world in the early eras of Christianity were his doctrines of God and humankind. Granted that even the *Dialogues* may be heavily edited and amended, and that Platonically-informed Christianity was heavily affected by neo-Platonism, the hierarchical but fluid dynamic structure of the milieu of the early church was indeed heavily dependent upon Platonic origins. In contradistinction to the Christian doctrine of *creatio ex nihilo*, in which God remains strictly transcendent over the world, Plato believed that God created the world through secondary powers or causes.

In analogy to the Babylonian myth of creation, such causes create by regulating order out of disorder, thus creating the cosmos out of chaos.[26] In that order belongs to the nature of divinity, God not only penetrates nature, but the world itself is divine. Plato calls it "a *god*".[27] The world has a soul that is "interfused everywhere from the center to the circumference of heaven".[28] Thus, rather than God, the creator, living in transcendence over the world, he interpenetrates it with "divine" and "necessary" causes that order and reorder reality. As a consequence, the divine may be sought in all things. The *things* of natural reality are secondary to the divine essences of reality which are responsible for the continued existence of the *things* themselves.[29]

Accordingly, all things partake of being and unity.[30] In that existence is distributed in things both the *one* and the *many* are expressions of unity for "every part is in the whole".[31] "The united being is one and many, whole and parts, limited and yet infinite in number".[32] Likewise, whereas on one level "the other" (the irregular motion of the finite) is opposed to "the same"(the circular motion of the infinite), the difference is only one of degree. Both are essentially in "the same state and that which is in the same state is like".[33] To illustrate this, let us take a look at how Nicholas of Cusa (c.1401 - 1464) emphasized this point. On the basis of Parmenides,[34] he argued that, whether *the united* partakes of a figure, be it "either rectilinear or circular, or a union of the two" [impossible as such a figure would be in actuality] "the middle will be equidistant from the extremes".[35]

God, according to Plato, is perfect in himself. He is "mind", or "the first principle". He is not only responsible for the creation of the soul, the world, motion, and the generation of all things, he is *reflected in them*.[36] As God moves earthward from his place of transcendence, he is met from below by the members of humankind who, inspired by God, reciprocally reach upward to fulfill themselves in the desire to become divine. The scheme, repeated again and again in Plotinus, Proclus and the Hermetic writings, sets forth the whole of reality as an essential harmony that has its real existence *in* God. "The ruler of the universe has ordered all things with a view to the preservation and perfection of the whole and each part has an appointed state of action and passion: and the smallest action or passion of any part affecting the minutest fraction has a presiding minister".[37] The "presiding ministers", the gods, who are subordinate to Plato's "God", "mind", or "first principle", affect the world's movements, generations and change. They are the divine powers which at one and the same time transcend nature, but nevertheless interpenetrate and control it. They are the metaphysical *causes* behind things.[38]

It is exactly this dual aspect of God, God in his aseity above the world and yet God intermingling with the world, that causes these theologies to be easy prey to pantheism, panentheism and mythology. Here, of course, is the basis for K. C. F. Krause's (1781-1832) panentheistic attempt to reconcile the platonically-inspired philosophies of Immanuel Kant (1724-1804), J. G. Fichte (1762-1814), F. W. J. von Schelling (1775-1854), and G. W. F. Hegel, (1770-1831), be it the subjective or absolute idealism.[39] Here too and even more directly are the roots of Whitehead's God of primordial and consequent natures. This is also the reason why the "process theologians" have difficulty understanding the "radical transcendence" of God and why their confusion of

God with the world inevitably ends where it begins, in one kind of pantheistic or panentheistic scheme or other.⁴⁰

It was not without significance, therefore, that Plato castigated those philosophers of nature (the pre-Socratic Ionians) who refused to recognize "gods" behind the elements of fire, water, earth, and air. They maintained that these moved by chance "and not by the action of mind, as they say, or of any other god, or from art, but . . . by nature and chance only".⁴¹ Plato was serious about the matter and this can be seen in the measures he set out to deal with those who refused to understand the world as he defined it. He argued that the impious, *i.e.*, those who denied that divinities interpenetrated and controlled the world and who refused them due honor, should be imprisoned. Those who were convicted for want of understanding rather than malice were to be sentenced to the house of reformation, for a period of imprisonment of "not less than five years". During their incarceration, the misguided were to be offered instruction that was intended to heal their souls and reform their thoughts. However, for Plato, as for many who have followed in the train of Greek philosophy, tolerance lasted only so long as there was confidence that his goals of reform would be achieved. When the time of imprisonment had expired, Plato advised that,

> if any of them be of sound mind [i.e. if they repented of their heretical ideas] let him be restored to sane company, but if not, and if he be condemned a second time, let him be punished with death.⁴²

Thus Plato was aware of the great chain of being that ran from the ruler of the universe through the minor gods or intelligences behind nature to the minutest part of nature itself. All was directed by God and toward the common good, each part was executed for the sake of the whole and the whole for the sake of each part. Harmony reigned. What was best for the individual is best for the universe. All *things* were subject to change and things changed according to their souls. Lesser changes maintained the soul at level ground. Changes that led to greater crime sunk it into the abyss or Hades. However, the soul that had *communion* with the divine virtue became itself divine. It was "carried into another and better place which is also divine and perfect in holiness".[43]

Plato's *spiritualization of reality* served to explain both the apparent erratic behaviour of the heavenly bodies and tergiversating occurrences on earth at a time when no other explanation was forthcoming. The whole of reality thus stood under the aegis of divine and mystical powers. The cosmos was understood as an animated organism with each part breathing life and directing itself to its god-determined, self-determined end.

Although nature for Plato was secondary, a manifestation of the divine and the real, in that it reflected the divine Plato displayed a profound interest in understanding the order of nature. Thus, while Plato eschewed any type of experimental science, even that of observation, he does not dismiss the "arts", as the "sciences" were called, altogether. In *Philebus*, for instance, he admitted that those sciences or "arts in which arithmetic, mensuration, and weighing" are used are also useful for knowledge.[44] However, even "the knowledge at which geometry aims" is, for Plato, "of the eternal, and not of the perishing and the transient".[45] Plato's interest in the philosophical and theoretical rather than observational, much less the experimental, is reflected in the *Philebus* where Socrates insists that the truer science has to do with "the permanent and imperishable and everlasting and immutable", as over against that which has to do with "the transient and the perishing.[46] It is in this light that, in the *Republic*, Plato recommends that mathematics, astronomy and harmonics are to be taught as *theory* without reference to astronomical observations or to acoustical experiments.

Even in his later so-called "scientifically-oriented works", Plato continues his emphasis upon the eternal and eschews that which is temporal.[47] In the *Timaeus*, for example, Timaeus, the Pythagorean, enters into a discussion of cosmology which prefigures Aristotle's *De Caelo*.[48] Even in this "most scientific" of his treatises, Plato continues to champion the theory of forms, of timeless and immutable objects, over the perishable things that were presented by the power of the senses.[49]

So, too, in the *Theatetus* Socrates argues that knowledge and thinking cannot rest upon sensations, because, as Descartes, too, was to believe, the former (thought) is permanent and the latter (sensations) are only transitory.[50]

Thus, according to Socrates, perception "cannot have any part in science or knowledge".[51] We are, therefore, advised to seek knowledge not in perception but "in that other process . . . in which the mind is alone and engaged with being".[52] Despite the fact that in the *Timaeus* Plato reports certain astronomical observations and refers to the data of anatomical dissections, he does not move in the direction of *observational science*. For him, as for Descartes some 2,000 years later, true knowledge cannot result from empirical knowledge. Knowledge is purely a matter of the mind or, better, a matter of "the soul by herself" seeking beauty and essence.[53] In a strikingly similar way, Descartes continues the tradition of metaphysical dualism when he proclaims *cogito ergo sum* (I think therefore I am). Thereby he provides the foundation of modern metaphysics.

From the foregoing it would seem possible to say that the actual contributions Plato made to the development of science were, first, his proclamation of an orderly world and, second, his insistence on the study of mathematics.[54] There are, however, two other contributions of Platonism that, although difficult to measure, were immensely important for the milieu in which science was to develop after the Renaissance. The first was a type of *via negativa*. Its contributions were both negative and positive. In refusing to allow the phenomenal world the same status as the *real world* (the unchanging forms constituted the *real* behind the changing and, therefore, *phenomenal* reality was less than real), basic and reliable knowledge of the things of this world was quite out of the question. This is its negative contribution. Accordingly all descriptions of reality, including those of the medieval and early Renaissance Aristotelians, were *ipso facto* inaccurate and inadequate in relation to the *real*. They could be, and indeed they were relegated to *speculation*, because they refered to the way things were seen. They were thus to be regarded with a certain degree of scepticism and as being of questionable value at best.

The second contribution of Platonism in this context was the characterization of reality as mystery, the depth of which could not be fathomed by any kind of human rationality. Certainly no kind of deductive rationality upon which Aristotelianism was based could comprehend reality. The conviction that reality was essentially mysterious and unfathomable led in the course of things to pure flights of irresponsible fancy. On the other hand, it is exactly the conviction that reality was essentially mysterious that both saves science or any other thought structure from becoming an all-deductive closed-world system. Science is driven forward in its attempts to penetrate the depths of reality that were and will always be beyond the reach of both the insight and wisdom of any and all epistemological inquiry. Hence, science is at base a never-ending pursuit of the partially knowable. Discovery depends upon the utilization of both intuition and imagination that reach beyond the presently known and presently rational into the not yet known and for which ever new rational processes may need to be discovered.[55]

In that there is much in Plato that stresses the mystery and the transcendence of reality, it is no coincidence that the Plato known to the early Christian church was the Plato who stressed the mystical heavenly harmonious world, the world of heavenly divinities, a world of souls that intermingled with the bodies and other material things of earth.[56]

Plotinus, Neo-Platonist
The Platonic analysis of reality and its attitude toward it was transmitted to and through the Middle Ages by the Augustinian neo-Platonism that was originally rooted in the neo-Platonism of Plotinus. The neo-Platonism espoused by Plotinus was itself a combination of Platonism and Aristotelianism. From Plato Plotinus inherited a primary interest in the "Good" or the "One", "Being", the "Soul",

"Beauty", "Love", and "Evil". These ideas pertained to "emanation", and the "descent of the soul into the body".[57] From Aristotle he inherited his interest in the "Nous" as the uppermost deity, as intelligence, and as the intelligible part of the soul. His ideas on "generation and order", the understanding of "existence" as "potential" and "actual", "substance" as consisting of "qualities", and the substantiated "unity" of all things were also derived from Aristotle. These were differentiated into concepts of "circular movement", "eternity and time", and the "animate and the man".[58] As for Plato and Aristotle, Plotinus insisted that "form" was primary to "matter" and that the "intellectual" was primary to the "sensible."[59]

Strict differentiation between Platonic and Aristotelian ideas in Plotinus' thought is, of course, difficult if not impossible because Aristotle had much in common with his teacher. Plotinus inherited the attempts of such Platonic scholars as Albinus (fl. 2nd cent. A.D.) and Apuleius (fl. 2nd. cent. A.D.) to synthesize the two. Basic to both Plato and Aristotle was the concept of an underlying harmonious relationship of all things. This fundamental conception of harmony penetrated the whole of reality and so affected the mind-set of the later neo-Platonists: Augustine, Proclus, Pseudo-Dionysius and the Hermeticists that it was to become the *sine qua non* of, and absolutely primary to, middle and late Renaissance thought. Reality was totally dependent upon the *One* who was *all*. Inter-related beings were arranged in an hierarchial order. These had inbuilt souls that were in constant movement downward from *the One* of their origin to the realm of *the below*. From *the below* they strove to move upward to *the above* where they would return again to their original pristine state of unity.

The thought of Plotinus that has come down to us has been brought together in the *Enneads*. Plotinus is known to have conducted seminars or discussion-like classes (*synousiai*) in which some of the participants took notes (*scholia*). It would seem legitimate to speculate that, as with the case of Karl Barth's (1886-1968) *Homiletik,* or Dietrich Bonhoeffer's *Christologie,* the *Enneads* are composed of student notes taken from the aphoristic statements Plotinus made somewhat at random. The arrangement and the organization of his writings into six parts of nine treatises each (*Enneads*, "The Nines") was done by his devoted friend Porphyry (c.232 - c.301). Porphyry described his relationship to Plotinus by saying, "I myself, Porphyry of Tyre, was one of Plotinus' very closest friends, and it was to me that he entrusted the task of revising his writings".[60]

It is also from Porphyry that we learn that Plotinus wrote at the time when the Christians "abandoned the old philosophy", *i.e.*, Platonism.[61] Porphyry apparently attempted to rectify matters by editing Plotinus' thought in the *Enneads* as a neo-Platonic apologetic in the face of Christianity. Porphyry classifies Plotinus as a "Platonist" along with Origen, both of whom were students of Ammonius Saccas (c.175 - c.243). They were, he said, "men greatly surpassing their contemporaries in mental power".[62] Porphyry, who was not given to understatement with regard to the admiration he had for his teacher, states that Plotinus understood both the Pythagoreans and Plato better than anyone prior to his time.

In addition, he assures us that, "Aristotle's Metaphysic, especially is condensed in them [Plotinus' writings] all but entire".[63] Perhaps with a view to establishing his master's reputation as a universal scholar, Porphyry asserts that, in addition to the doctrines of Plato and the Peripatetics, Plotinus included the doctrines of the Stoics within his writings. In addition Plotinus had a thorough knowledge of geometry, mechanics, optics and music. He cautions, however, saying, "It was not in his [Plotinus'] temperament to go practically into these subjects."[64] Nevertheless, Porphyry assures us that Plotinus' sources did not determine his course. "He followed his own path rather than that of tradition".[65]

Porphyry's adulation for his master Plotinus was genuine. His account of Plotinus' appearance in a divine form follows an intellectual norm of his era. He tells us that an Egyptian priest once evoked the visible manifestation of Plotinus' presiding spirit in the temple of Isis, *the only place which the*

Egyptian could find pure in Rome. "At the summons a Divinity appeared . . . and the Egyptian exclaimed, 'You are singularly graced: the guiding-spirit within you is none of lower degree, but a God'". Porphyry, in a proper neo-Platonic mood, then goes on to explain, "thus Plotinus had for indwelling spirit, a Being of the more divine degree, and he kept his own divine spirit unceasingly intent upon that inner presence".[66] The statement is surprisingly close to Paul in the Letter to the Colossians, "For in him the whole fullness of deity dwells bodily". (Col. 2:9)

The *Enneads* are centred on *The One*. *The One*, for Plotinus, is the source of reality. From it all else emanates. All depends upon *the One* for its existence and since, in the end, all will return to *the One*, even now all aspires to return to its source. The life of the soul is *motion*, and motion defines the life of the cosmos that itself is beautiful. The divine, the One, or the soul, is eternal and undivided as over against the "material" that consists of "bodies" and the "bodies" are many, changeable and transient. Consequently language, being of the material world is, of course, not adequate to God, the supreme Being. True knowledge of the One is, therefore, inexpressible. Thus the soul that knows the One does not know the One intellectually. Rather, it knows the One by becoming the One. Such a soul is "caught away, filled with God".[67] At the same time, and very important for later neo-Platonic development: all participate in the Good. All souls are one soul. The inbuilt symmetry of the universe includes all its separate parts.

Three basic hypostases (natures) form the trinity of *the One, the Intellect*, and *the Soul*. They are in constant animated interaction. In the end and by mystical exaltation, there is a return of all to the One. This movement is "from nothing alien . . . to nothing alien".[68] In that each soul is a "member of the divine" by virtue of "the Intellectual-Principle" that exists within it, the proper vision of God and of ourselves "is of a self-wrought to splendour, brimmed with the Intellectual light, become that very light, pure, buoyant, unburdened, raised to Godhood or, better, knowing its Godhood all aflame then". Such a state is "the knowing of the self restored to its purity".⁶⁹

Plotinus' conceptuality presages Paul Tillich's (1886-1965) designation of God as: "the ground of being", "being itself", "the power of being", and "the absolute itself that is the ground of all absolutes", as well as his understanding of immortality as "essentialization".⁷⁰ He speaks of the essential man as outgrowing Being and becoming "identical with the Transcendent of Being". He strives for "self principle", the "unity", "unity absolute", "the Good that is above all that is Good". The home of the soul is in God and all strive to take their proper places in the realm of the divine. The proper end of human beings is to be "brought back to their origins, lifted once more toward the Supreme and One and First". In reaching toward this goal in which unity is attained, the soul "puts away the evil of earth, once more seeks the Father and finds her peace".⁷¹ Augustine's, "Our hearts are restless until they rest in thee", has obvious precedent.⁷²

In contrast to this "life of gods and of the godlike and blessed among men", there is "the alien that besets us here, a life taking no pleasure of the things of earth". From this alien state the soul must be liberated. Even the alien things of earth, things for which, as Porphyry explains, Plotinus had not the temperament, are not consigned to unreality or non-existence. In that "everything has something of the Good by virtue of possessing a certain degree of unity and a certain degree of Existence and by participation in Ideal-Form" neither matter nor the world is evil as such.⁷³

Hence against those who affirm that the creator of the cosmos and the cosmos itself is evil⁷⁴ Plotinus affirms that "the same Nature [that belongs to things] belongs to the Principle we know as The One" which itself is good. Hence, we are warned that "to despise the [earthly] Sphere, and the Gods within it or anything else that is lovely, is not the way of goodness". Plotinus is not consistent, however. He can also say, "In a word, life in a body is of itself an evil", and even the soul that accompanies the body "enters its Good through Virtue . . . [and] holds itself apart".⁷⁵

So, too, in the animal and vegetable realms, there is no

evidence that "the Reason-Principles", understood in this context as principles of form are identified with matter. Matter in itself is dead and the soul communicates nothing at all to that which underlies it. So distant are soul and body from one another that, in spite of the fact that the souls are interspersed in bodies, there are no thought categories which are common to the realm of the "divine", on the one hand, and to the region of the "material", on the other. The divine and the One, or soul, are all eternal. Material or matter is divided, changeable and transient, an entity of ceaseless flux. The soul is the "principle of permanence". Bodies, therefore, consist of a combination of matter and soul. The matter is the material aspect of the body; the soul provides the shape or form.[76]

While the universe for Plotinus "in its material mass, has existed forever and will forever endure" there is constant flow between the "spiritual" and the "material".

> The will of God itself is able to cope with the ceaseless flux and escape of body stuff [which occurs in death and decay] by ceaselessly reintroducing the known forms in new substances, thus ensuring perpetuity not to the particular item but to the unity of the idea.[77]

Permanence is a problem. Things of earth, *i.e.*, things below the orb of the moon "have merely type-persistence" while the "celestial realm and all its several members possess individual eternity".[78]

Although, as we have seen, Plotinus' basic Platonism dictates that matter is dead, in an Aristotelian kind of way, matter, for Plotinus, has a degree of worth and is caught up in ceaseless flux with the divine. It constitutes the physical mass of the universe which strives "towards the immortality of the cosmos". Not that matter is autonomous. A soul hovers over all matter and every soul circles about the godhead in its own rank and place. It is the lowest power of the soul that is nearest the earth and is woven throughout the entire universe. One phase possesses sensation while another includes reason and is concerned with the objects of sensation. The one hovers above the other, "giving forth to it an effluence which makes it more intensely vital".[79]

In Plotinus' hierarchical world, therefore, every individual and every thing is under the power of its "soul". The souls are imprisoned in brute bodies to be sure. Even human souls find themselves in "bitter and miserable durance in body, a victim to troubles and desires and fears and all forms of evil, the body its prison or its tomb, the cosmos its cave or cavern". In themselves, however, souls are immortal. Thus, in no sense does this imprisonment or entombment qualify the soul as such. Its task is intellection.

> It looks toward its higher and has intellection; towards itself and conserves its particular being; towards its lower [which it] ... orders, administers, and governs.[80]

As a matter of fact "through soul this universe is a God: and the sun is a God because it is ensouled; so too the stars; and whatever we ourselves may be, it is all by virtue of soul; for 'dead is viler than dung'". Hence our own natures are the same "ideal nature", like the "oldest God" of them all. The One that is above the multiple god (the cosmos) is "essentially a unity untouched by the multiple...[and this unity] everyone desires to penetrate". However, "only by a leap can we reach this One which is to be pure of all else". The move toward the One is a search for unity. At one and the same time, the unity is separate from all things; but in it all things participate.[81] To search for this unity is to seek the "principle of all, the Good and the First".

> We must strike for those Firsts, rising from things of sense which are the lasts. Cleared of all evil in our intentions toward The Good, we must ascend to the Principle within ourselves; from many, we must become one; only so do we attain to knowledge of that which is Principle and Unity.[82]

"To think of the One as Mind or as God" is to think too meanly. It is rather a "self-sufficing principle", a "Unity-Absolute". It is the Good, "the Good above all that is good". It is thus a type or an anti-type of "ultimate reality" or "the absolute itself, the ground of all absolutes" that one finds repeated in Tillich.[83] Reaching toward its goal of unity, the soul "puts away the evil of earth, once more seeks the Father and finds her peace". To see the self properly, then, is to see the self as belonging to the order of the Godhead and merged into it. It is the knowing of the self restored to its purity. The self thus lifted, is in the likeness of the Supreme. "If from that heightened self the soul should pass still higher — "image to archetype" — the goal of all journeying is won. If by chance there should be a falling back, the journey begins again.

> We waken the virtue within until we know ourselves all order once more; once more we are lightened of the burden and move by virtue towards Intellectual-Principle and through the Wisdom in That to the Supreme.[84]

Surrounding all is the Intellectual-Principle (Aristotle's prime mover) which has "no progress in any region. Rather, "its movement is a stationary act, for it turns upon itself... circling as it does, [it] is at the same time at rest".[85] This turning in perfect circularity is of the nature of perfect motion and, therefore, of the heavens in their entirety. Strangely, or perhaps not so strangely, in a passage which can be seen as a direct precedent of Nicholas of Cusa's *De Docta Ignorantia* wherein the world is so constructed that it has "its center everywhere and its periphery nowhere", for Plotinus "every star, considered in itself, is at center with regard to some one given group and in decline with regard to another and vice versa".[86] The whole interrelated world is such that:

> Each entity takes its origin from one Principle, and, therefore while executing its own function works in with every other member of that. All from which its distinct task has by no means cut it off: each performs its act, each receives something from the others, everyone at its own moment bringing its touch of sweet or bitter. And there is nothing undesigned, nothing of chance, in all the process: all is one scheme of differentiation starting with the Firsts and working itself out in a continuous progression of Kinds.[87]

Thus, "All living things — all in the heavens and all elsewhere—fall under the general Reason-Principle of the All — they have been made parts with a view to the whole".[88]

Such was the basis of the interrelated, convoluted, harmonious, soul-filled, divine and unity-directed world of neo-Platonism with its circular turnings, its spirals of souls from heaven to earth and returning back again to the Supreme. Such holy secrets are not for the uninitiated. Such pearls are not to be thrown before the swine. "The holy things may not be uncovered to the stranger, to any that has not himself attained to see."[89]

The Systematic Theology of Proclus
Four centuries after Plotinus, Proclus attempted to systematize neo-Platonic thought. This was one last attempt to protect the ancient world from the inroads of the vulgar religion of Christianity. Nowhere in the whole of antiquity are we as fortunate in discovering the essence of neo-Platonic thought as in the philosophy and theology of Proclus' *Elements of Theology*. Proclus, who is also known as "Diodochus" (the follower) because he succeeded to the head of the Platonic Academy in Athens, was also a follower of Plotinus. Writing in the sixth century this "last of the Greek philosophers" combined the teaching of the *Chaldean Oracles* with the doctrines of Plato's *Timaeus*. This was the first and only systematic treatment of the neo-Platonic system of metaphysics that we know to be extant. The system is of more than of passing interest. When Pseudo-Dionysius "Christianized" or "baptized" the thought of Proclus into the Christian faith, he used ideas in the *Elements of Theology* to develop his system of celestial hierarchies. The Dionysian system had such a powerful hold on the medieval mind that it became the basis for the piety of the medieval church.

While Porphyry tells us that Plotinus had a proper understanding of both Plato and Aristotle, Proclus contends in his commentaries on Plato's *Dialogues* that he will be more faithful to Plato than was Plotinus. "It is requisite", he says, "that we should be persuaded by what Plato has demonstrated, and by the most efficacious attestation given by the [Chaldean] oracles to the demonstrations of Plato".[90]

Considering the role Hermeticism was to play from the Renaissance on, it is interesting to note that after recognizing the authority of Plato and the *Chaldean Oracles* Proclus also pointed to the Hermetic literature as a source of his thought. He recognized Hermes as "the common leader" regarding questions of the soul.[91] Plato, however, is the philosopher of philosophers. Compared to Plato the rest of Greek philosophy, Proclus tells us, was useful only for interpretation. So dutiful was he to Plato, in intention at least, that, rather than fault him for the inconsistencies of his writings, he excused Plato's apparent discrepancies as resulting from the crudity of the reader's interpretation. Proclus was so concerned that the reader might easily get it wrong that, like the ancient Pythagoreans and like Plotinus also, he restricted the circulation of the texts to those who were firm enough in the neo-Platonic faith not to be injured by them.[92]

For Proclus, as for Plotinus, the whole of reality, universals, and particulars are dependent upon *the One*. The One exists above all things. Providence exists in accordance to the One. Referring to Plato's *Phaedrus* Proclus adds "*the Good* is the same as *the One*". It illuminates all things and conserves them, arranges them and turns them back to itself and "imparts good to all things". *The One* is the source of *the Good*.[93]

According to Proclus all things participate of "the One". There is, thus, a fundamental unity of all things. The universe is hierarchically arranged from Being through Intelligences, Souls, Forms and Corporate Bodies — all separate yet all inter-related. As in Plotinus, so in Proclus, *matter* is essential to corporate bodies; and although like all things it issues from *the One* and participates *of the One*, it is the lowest of "existents".[94] Beyond all bodies is the soul's essence; beyond all souls, the intellective principle; and beyond all intellective substances, the One.[95] Further, behind, but at the same time integral to each body, is a *soul* and beyond each soul, an *intellectual principle*, and beyond all the intelligences, *the One*. The souls, directed by the intelligences, strive toward oneness and unity both with regard to their natures and in relation to their positions. If they are *composite*, they *descend*, and if *simple*, and thus "near to the One", they *ascend*. Thus each soul's destiny, its descent or ascent, depends upon the encumbrances of its composition. Souls that choose the appropriate life and become *simple* return to their origin in the One. There they are reunited with the unity from whence they came.⁹⁶

Proclus' *Elements of Theology* is not as edifying as the *Enneads*, nor is the work a complete epitome of neo-Platonism. It omits, for instance, the whole subject of cosmology. Nonetheless it does present a succinct and reliable understanding of the kind of neo-Platonic philosophy that helped shape both medieval piety and the Renaissance mind. Proclus begins his theology with the problem of "the One and the Many". He proceeds as does Plato in the *Laws*, and indeed as does all "natural theology", from the phenomenal world in which it finds evidences of God. Plato begins with the world of existence and proceeds to "the One", "the Good", or what Tillich would call "the ground of Being".⁹⁷

Proclus' theology is both systematic and in its own way, logical. It attempts to move from the "simple" to the "complex" in order to demonstrate that *all* participate in the *One*. The *One* is also *Unity* and *Good*. He begins with the "ones" of the world of experience. From the "One" he moves to the "many". As found in the world of phenomena the ones are not pure because purity belongs to *Unity* which is absolute. The ones, nevertheless, participate in the *One* which causes them. Things constituted of ones are manifolds and the individual units of the manifolds all participate in the *One*.⁹⁸

Next, under the influence of Aristotle, Proclus explains "causes". Causes are "transcendent". In that cause is superior to effect, everything effected must have been caused by that which is superior to itself. Unity and Good are first identified with both the "Final Cause" and the "Efficient Cause". The Good, then, is identified with the *One*.⁹⁹ And the different grades of reality that run from below to above are dependent upon their proximity to the One. There are "bodies" that are moved, and "Souls" incorporeal and independent of bodies that are "self-moved" but responsible for the apparent self-movement of bodies. The self-movement of bodies is thus dependent upon the presence of life or Soul. The Soul in turn participates in "Intelligence" which is itself subordinate to the *One*.¹⁰⁰

Although the *order of knowledge* for Proclus proceeds from the plethora of bodies or ones to *the One* in a strictly upward direction, the *ontic order* is circular. It proceeds by "emanation" from the top downward and also by "reversion" from the bottom upward. The result is that, just as the *One* is immanent in all reality, so by means of a reverse movement, all reality moves back again into the *One*.¹⁰¹

"Time" and "Eternity" are principles that qualify realities. Realities are "self-constituted." They are without temporal origin and, therefore, they are "imperishable" and "perpetual". Contrarily, "that which is measured by time" is either "in existence or in the process of coming to be". While that which is eternal is "whole", that which exists under the exegencies of the "temporal" is "multiple". However, since each *one* is part of the *whole*, all things, even those which exist under temporal conditions continue to participate in the eternal.¹⁰²

"Wholeness" for Proclus is metaphysically situated between "Being" and "Form". It is an *intermediary* between the two. Being is prior to Form. Through wholeness Forms, which give existent things their particularities, make it possible for things to participate in Being. All things that exist are, therefore, "in some sense existent" because they have received "some feeble irradiation of existence" through the "unitary power of being".¹⁰³

In reliance on both Plato's *Phaedo* and *Timaeus* as well as Aristotle's *Metaphysics,* Proclus speaks of the "unmoved cause and moved cause" much as Plato spoke of "prime causes" and "second causes" and Aristotle spoke of "motion" and "the source of motion".¹⁰⁴ Accordingly all causes proceed from the "first principle of all things". The "incorporeal acts upon the corporeals", even as the former participate in the latter through "an inseparable potency which it implants". This potency makes the corporeal capable of "self-reversion", so that all that 'perpetually is, is infinite in potency".¹⁰⁵ All that exists depends upon "Primal Being". All that has "self-movement" lives because of "Primal Life". All that is cognizant participates in knowledge because of "Primal Intelligence". The conclusion of this argument was very important for both ancient and Renaissance neo-Platonism. It claimed, "all things are in all things" (πάντα ἐν πᾶσιν). Hence, all things are a single reality distributed in a hierarchy of being which proceeds from Being itself to "corporate bodies" through "Intelligences" to "Souls" and "Forms" and back again by way of reversion.[106]

Added to all this are the "divine Henads" in which the Divine is hypostatized. All participate in the "divine Henads" or "gods" which are themselves numbered but which also participate in a divine Unity. Every god is above life, and above intelligence. Every god is participative, except the One, the "First Principle", or the "Good" from which all goodness, unity and excellence derive.

The "Divine", in itself ineffable and immovable, "exercises providence toward secondary existence which it transcends". It may be apprehended and known from these existences since they participate in it. Only the First Principle is "completely unknowable". The gods, reveal themselves in their manifestations by proceeding through secondary order, multiplying, bestowing, and yet preserving the distinctive character of their nature in the process. Divine bodies receive their divinity from above. Divine bodies are divine because of the "mediation of the divinized soul". Divine souls participate in "divine intelligences" and the divine intelligences participate in a "divine Henad". The Henad is the "immediate deity", the intelligence "most divine", and the soul "divine".[107]

Thus, while the "primal God" is "the Good unqualified" and "Unity unqualified" each of the gods which are posterior to the primal God is a "particular Excellence" and "particular Henad". The more universal the god or the Henad, the closer it stands to the One. In every divine Henad one "real existent" participates. Whatever is divinized is "linked" by "upward tension to one divine Henad". Thus, "every Henad is cooperative with the One in producing the real-existent which participates in it". The sequence of principles that participates in the divine Henad extends from being to bodily nature. The highest is Being and the lowest is the body. It is "the last participant".¹⁰⁸

The gods, who take their origin from above, proceed via intermediaries to the "last existents and the terrestrial regions". Thus, they are "present alike to all things". The procession of both "things existent" and "orders of existents" extends just as far as "the orders of the gods".¹⁰⁹ However, "not all things are present alike to the gods". There are "divine ranks" and "divine processions" and the clue to the movement of the whole heavenly hierarchies is that, "In any divine procession the end is assimilated to the beginning, maintaining, by its reversion thither, a circle without beginning and without end."¹¹⁰

Like all being so, too, the ranks of the gods are interconnected. They are "finite in number". That which is "paternal in the gods" stands at the head of the ranks. Their "generative" capacity results from the "infinitude of divine potency". In addition, there are attributes of "perfection", "protection", "zoogamic" or "life-giving", and "purity". It is the function of the "paternal causes" to bestow being to all things and give original "substantive existence to all that is". Somewhat unspecified are the "evaluative causes" which "liberate from the lower principles". Even they are "found primitively in the gods".¹¹¹

The orders of the gods derive from the principles of "Limit and Infinity". Both are present but sometimes the one dominates and sometimes the other. The Henads participate in "Divine Intelligence", "Perfect Unity", "True Being". The gods are named from the principles which are attached to them.¹¹² In addition the Henads serve to bridge the gap between the One and the Forms. They may be either supramundane or intramundane. If supramundane, the soul "enjoys participation" in them. If intramundane, any sensible body participates in them. The "intelligences" stand between "the One" and "the minimal unity of matter". By acts of intellection, intelligences give rise to the "intellectual forms". These in turn are constitutive of the things that are "perpetual" and are "participated by souls".¹¹³

"The Souls", which are next in the line of descent after the Henads, are either divine, in which case they are spiritually indivisible, or they are subject to change. If qualified by their association with the temporal being, the Souls are differentiated in relationship to bodies. In that case, while maintaining "eternal existence" they also engage in "temporal activity". Their movement in time, which has "perpetuity", is measured in "periods" as well as in "cyclic reinstatements". For "in the case of things perpetual, every period ends in a restatement of the original condition".[114]

In that every Soul bears "a vehicle of the divine soul" in respect to its being, it "can descend into the temporal process and ascend from the process to Being an infinite number of times".[115] Having been created by "an unmoved cause", the "vehicle" is "immaterial", "indiscernible", and "impassible". Every vehicle of the soul " descends by adding vestures increasingly material and by acquiring the irrational principles of life". Divesting itself of its material, it ascends and recovers its proper form, "putting off all those faculties tending to temporal process . . . and becoming clean and bare of all such facilities as serve the uses of the process". Souls vary in shape and size perpetually because of the addition or removal of other bodies, but they ascend or descend in their entirety.[116]

Thus, the final and only systematic presentation of neo-Platonic theology of which we have knowledge, presents us the ligaments of that mysterious universe that begins with the One, the Being and the Good and proceeds by perpetuating "that which is primary throughout reality through the mediation of gods". Intelligences, Forms and Souls descend and as they take on the material and the irrational, they enter the temporal world. Then, in a cyclic movement of recession, they ascend by divesting themselves of all that they have accumulated in the temporal-material world. They thereby cleanse themselves of all that is foreign. Thus, they regain the rational, eternal reality of the Being and the Good from whence they came. The cyclic downward and upward movement is perpetual.

The whole mysterious, inter-related, cyclic, dynamic order was, as we have already said, "Christianized" by Pseudo-Dionysius. In the *Summa Theologia*, Thomas Aquinas used the myriads of Pseudo-Dionysian angelic beings to transfer holiness from heaven to earth in a descending and ascending pattern. In a burst of ecstatic imagery, Dante reinforced the scheme. Angels spiral from heaven to earth and back again. Those who are "participated" (to use Proclus' terms) in by them are blessed with bliss on earth, even as they await the time of "recession" to heaven. When their souls put off their earthly accoutrements, they ascend back to their heavenly home. There they repose eternally in the restfulness of participation in God who is their author and in whom alone is perfection, unity and peace.

Hermeticism's Cosmic Cult
Proclus' reference to "Hermes", cited above, raises the subject of yet another, certainly the most complicated and, in some way, the most important of the neo-Platonic writings to reach the Renaissance. These are the writings of Hermes Trismegistus. These writings originated in the second and third centuries A.D. and purport to be written by Hermes, an Egyptian priest who was somewhat younger than Moses and just older than Plato.

The Hermetic writings or the *Hermetica*, as they are now referred to, combine neo-Platonic mystical elements, neo-Pythagorean numerology, philosophical concepts of the *Chaldean Oracles*, astrological aspects of Plato and Aristotle, and ancient alchemistic and cabbalistic concepts. In addition, there are just enough references to ancient Egyptian cultic practices to support their claim to have originated in ancient Egypt. The esoteric, mystical, magical, pseudonymous system was reputed to have had its origin with Hermes Trismegistus, (Hermes, the thrice-great), also called "the Egyptian Moses". He was even supposed to have been the source of the wisdom that Plato recorded in his *Dialogues*.

Hermes Trismegistus was not a stranger to East or West. He had been known and proclaimed by such Church Fathers as Clement of Alexandria (c.150 - c.215), Tertullian (c.160 - c.230), Athenagoras (fl. 2nd cent.), Cyprian (200-258), Lactantius (c.240-c.320), Augustine (354-430), and Cyril of Alexandria (376-444). Clement records that he knew of forty-two books of Hermeticism.[117] Lactantius knew Hermes as the author of "many books"[118] and Cyril spoke of a collection of fifteen such books compiled in Athens.[119] If Augustine knew little Greek, his first-hand knowledge of the Hermeticism must have been confined to the *Asclepius*[120] A manuscript containing the first fourteen books of what we know as the *Corpus Hermeticum* was brought to Italy in 1463 at the behest of the renowned cultural benefactor, Cosimo di Medici. The books were considered to be of such great importance that at Cosimo's orders, Ficino translated them prior to completing the translation of the *Dialogues* of Plato which he already had in hand.[121]

The claim that the Hermetic materials were of ancient Egyptian origin remained unquestioned until 1614 when Isaac Casaubon (1559-1614) redated the writings to the second and third centuries A.D. The persistent allegation of antiquity added to their appeal during the Renaissance with its avid revival of classicism. Even after Casaubon redated the Hermetic writings, they continued to have an enormous influence. This indicates that the mystical chord they struck fell on receptive ears indeed. The ears, of course, had been taught to listen for such exotic tones by the neo-Platonic theology of Augustine as well as by the Pseudo-Dionysian thought reaffirmed by Thomas Aquinas.

The *Hermetica* reiterate themes of emanation from and participation in *The One* which we heard sung over and over again by both Plotinus and Proclus. To this, however, the writings add a heightened sense of cultic mystery that includes cosmology. The whole of reality is depicted as vacillating between eternity and time. All flows from the One into the world and back again to the One. All things are controlled by God through his governors, the planets. People are God-like and are, therefore, in the privileged position of being God's instruments in governing creation. Human beings are thus the priests of creation, a concept that, as we shall see, has immense import for the development of science. In addition, and of at least equal importance, as far as the development of science is concerned, the writings present us with a cosmology that is interesting not only because it links the heavenly and the earthly spheres but, even more so, because it orders the planets about the sun in an heliocentric pattern.

The *Asclepius*, extant only in its Latin version, presents us with a world that follows the neo-Platonic hierarchial system in expansive detail. The whole of reality is seen as a divinely governed unity with God at the top of a great chain of being reaching from the higher heavens above to the depths of the earth beneath. The *Asclepius* proclaims an astrological gospel in which the sun, moon and planets all play their parts. The heavenly bodies reflect the perfect and divine harmony of the heavenly spheres, and this harmony reaches from *above* to *below*.

As in both Plato and Aristotle, so in the *Asclepius*, the elements fire, water and earth[122] are all dependent upon the same final essence. Thus they move out of one another and into one another, dividing and reuniting, decomposing and composing, according to the causes behind them.[123] Also, as in Plato, for whom the heavens reflect harmony and rationality, so in the *Asclepius* humankind, who are born of heaven but set on earth, are rational *insofar* as they are able to reflect the ways of God. Through them and with them God governs the world he has made. And very importantly, as far as the development of science is concerned, in reflecting rationally in the midst of the irrational worldly sphere, people participate in the renewal of the world itself.[124]

The same themes appear in the *Pimander*, the first book of the *Corpus Hermeticum* that Ficino translated in 1463. As in Plotinus and Proclus, the *Pimander* identifies God, the world and the self.[125] God is the *Nous*. He exists *prior* to the world. By his Word, who is both his *Son* and *Light*, God confronts the darkness and out of darkness brings forth both the world and the material elements. The souls, however, emanate from God. In addition to bringing the world into being by his Word, the *Nous* engenders the *Nous-Demiurge* who is at one and the same time spirit and the *god of fire*. In a play on Plato's world-creator, the Nous-Demiurge formed the so-called "seven governors", the seven then-known heavenly spheres that circle around the earth in eternal orbits. In response to the will of the *Nous* they shape the destiny of humankind and the natural world.[126] The rotations of the heavens are what we would call the "intermediate cause" of earthly creation. Through them God acts and from them the influences move down to earth and fashion earth's creatures out of air, water and earth.[127] The Nous-Father himself bears humankind in his own image. Hence humankind is a being like the Father, and the Father set them over all his works. Since humankind participate in the same nature as the Father, they are in essence eternal although their bodies are mortal.[128]

Whereas humankind share the eternal essence of the heavenly bodies, their earthly bodies are subject to mutation and change. Eventually, as in Plotinus and Proclus, the corporeal aspects will be separated from the eternal essence. While the material elements will return to the source from which they came, the spiritual essence of humankind will regain its pristine purity by putting off all material elements, bodily desires and positions. Then, unencumbered, it will re-ascend to the sphere of God from whence it came. There it will reassume its divine status both by entering into God and by becoming God.[129]

Treatise V, "Hermes to His Son Tat, That God is at the Same time Invisible and Most Visible", also follows the theme of the divine interpenetration of all things that is basic to the neo-Platonists, Plotinus and Proclus. God is the "ever-existent".

> He makes manifest all else . . . He himself is hidden. He presents everything to us, but not himself, to our senses. Rather he manifests himself in all things and through all things.[130]

God presents himself to ourselves by penetrating all things.[131]

Further, following a pattern of thought that is as old as the Pythagoreans, but that is reflected in Plato and Aristotle as well, God, the maker of all, works through the sun. He is "the greatest of the gods of heaven" and "to him as their king all the other gods of heaven yield place".[132] God, the maker of all things, is "too great to have a name", "although hidden he is most manifest". "God is known by thought alone; nonetheless we are able to see him with our eyes"

> Having many bodies, there is nothing that he is not, for all the things that exist are he. It is because of this that all names are names of him, for all things come from him, their own Father. Because he is father of all, he has no name.[133]

The religious motif which had accompanied and often dominated "scientific" efforts from the ancient pre-Socratics onward makes for a complete integration of theology and cosmology in Hermetic thought. In Treatise XI, "Mind to Hermes", God is the unique creator. He alone provides the soul as "the soul is one and life is one and the material is one", so too "the world is one and the sun is one and the moon is one and the activity of the divine is one", even as God is one. "It is God alone who has created all things".[134] God, who is incorporeal, encloses all things in himself.[135] God is to be understood as containing all things, even the entire world, within himself just as thoughts that he thinks.[136] Further, in that "like cannot be known except by like", to know God is to become equal to God. The reader is therefore advised:

> Make yourself great until you reach a grandeur without measure. By a leap which frees you from all that is corporeal; lift yourself above all time and become eternal; then you will comprehend God.[137]

In the light of the heliocentric theory which focuses on the centrality of the sun in the solar system, it is of more than casual interest to note that over and over again the Hermetic documents follow Plotinus in insisting upon the importance of the sun. The sun is the manifestation of God and the begetter of the good as well as the human beings on earth. The sun sheds its light on all below even as it orders the heavens and establishes the beauty of the world. It is the foundation of the universe where all comes into being, "the nursery and feeder" of all terrestrial creatures.[138]

Treatise XVI of the *Corpus Hermeticum*, "Asclepius to King Ammon: Definitions", which is a part of the Turnebus Latin edition of 1554, sums up the Hermetic theological cosmology in a discourse that begins with an invocation:

> God, the master, creator, father and encompasser of the entire universe, who is the One who in that he the One is all and who in that he is all is One. Because the completeness of all being is one and he is in the One not that the One is itself of the double, but that the two together are but the one selfsame unity.[139]

The discourse then goes on to explain that while God can only be contemplated, the sun can be seen. The sun illuminates the entire universe. It benefits not only the heavens by shining on that which is above, it also penetrates to the lowest depths of the earth below.[140]

> For the sun is stationed in the *midst of the world*. It carries the world like a corona and, like a good driver, it assures the balance of the chariot of the world. It [the world] is attached to himself so that it does not run off in a disordered course. The reins are life, and soul, and spirit, and immortality and generation.[141]

In short, the intelligible world depends on God and the sun, that penetrates the intelligible and sensible world, receives from God its supply of creative and life-giving energy.[142]

> Around the sun gravitate the eight spheres, dependent on the sun are the fixed stars, the six spheres of the planets and the unique sphere (that of the moon) which encircles the earth.[143]

The spheres in their turn are each accompanied by corps of demons (celestial deities). In that the sun depends on God, the spheres on the sun, the demons on the spheres, and humankind on the demons, all things depend on and reach up to God.[144]

Thus, Asclepius reveals to King Ammon the secrets of the cosmos. The cosmos is dependent on God as its source. God works through the sun and its planets that, as demiurges, are arranged in an Aristarchean heliocentric pattern with the sun in the middle of the world surrounded by the eight spheres. These are the spheres of the fixed stars, and the six planets (the five known planets plus the earth and the sphere of the moon which turns around the earth even as it, with the other planets, circles the sun).

Each sphere is accompanied by a bevy of demons that control the ways of nature and humankind.

> God is the Father of all things, the sun is the demiurge and the cosmos is the instrument of the demiurge's activity. The heavens are governed by the intelligible substance and it governs the gods (the stars and the planets) and the demons, under the orders of the gods, govern humankind.[145]

Thus:

> God created all things for himself by his mediators and all things are parts of God. In that all things are parts of God, God is certainly all things. Therefore in creating all things God creates himself and it is quite impossible for him to stop creating for God can never cease to be. And by the same token since God can never cease to be, so also with his creative activity, it neither begins nor finishes.[146]

Treatise V ends in a hymn of praise that makes it understandable why the Hermetic materials could so easily inspire the Christian mind of the late Renaissance.

> With what shall I sing to thee? For the things thou hast made for the things thou hast not made. For the things thou hast revealed or for the things thou hast concealed. And for what reason shall I sing to thee? Do I belong to myself or have anything of my own? Am I other than thou? Because thou art all that I am, art all that I do and all that I say, because thou art all nothing exists outside of thee . . . Thou art all that has come to be, thou art thought in that thou thinkest. Thou art Father in that thou hast fashioned the world. God in that thou workest the good and maketh all things.[147]

Ficino, Renaissance Platonist

Marsilio Ficino was by all odds the most renowned Platonist of the Renaissance. In 1462, under the patronage of Cosimo di'Medici, Ficino founded the Florentine Academy. In 1463, Ficino, who was a monk and sometime canon of the Florence Cathedral began his translations of the first fourteen books of what we now know as the *Corpus Hermeticum*. Because of his own interest in the writings of the "ancient" Hermes Trismegistus, Cosimo had sent for the manuscripts and had them brought to Florence for his own edification.[148] After Ficino had translated the Hermetic materials, he then rendered Plato's *Dialogues* into Latin between the years 1467-69. He went on to translate the writings of Plotinus in 1484, and those of Proclus in 1497. More than any other, therefore, Ficino is directly responsible for having placed both Platonism and neo-Platonism, especially its mystical aspects, on the stage of the late Renaissance theater.[149]

It was Ficino, therefore, who re-introduced these Platonic and neo-Platonic ideas to the West just in time for them to fertilize the late Renaissance mind. At that time it was partly due to the kind of exhilarating Platonism introduced by the writings mentioned that Aristotelianism was being pushed aside. The challenge to Aristotle had already begun as the scientific efforts of the late twelfth and early thirteenth centuries began to take effect. Before the middle of the fifteenth century it was coming to be perceived as too restrictive, too inadequate, and also as self-contradictory.

For Platonism nature itself was impermanent and transient. Any description of nature, even if it carried the prestige of Aristotle, must necessarily be impermanent and transient as well. In addition, because it was characteristic of the study of nature in Greek thought to classify any understanding of nature as *speculation*, all answers were necessarily questionable. At the same time Platonism *per se* seemed too esoteric to deal with the real world, and it was little match for Aristotelian logic once the first principles on which it was based were accepted.

Consequently it was not because there is an inherent dialectic between Aristotelianism and Platonism that the two systems alternated in capturing the attention of the human mind. Just as Aristotelianism had replaced Augustinian neo-Platonism as a basis for both theological thought and thought about the world in the late medieval period, Aristotelianism lost its grip in the middle Renaissance. At this point the tables turned and the legacy of Plato again rose to ascendency. It was not only Platonism that began to reform the Renaissance mind, but more particularly, the mystery-laden thought of Plotinus, Proclus and the cultically empowered notions of Hermeticism.

Thus the Platonic legacy did not consist of the pristine Plato but of Plato who had been integrated into the Renaissance mind. Now he was accompanied by various schemes that were indigenous to the Renaissance in the latter part of the fifteenth century. In part it was the Plato of the Franciscan reformers, Grosseteste, Roger Bacon, Duns Scotus and William of Ockham. However, Platonism was presented to the middle Renaissance in its most powerful form by Ficino. This is especially important because incorporated into Renaissance thought in this fashion, it was to have a profound impact on the development of modern science.

Ficino, like a majority of the intelligentsia of his time, was an ordained priest. Although for the most part the church remained tied to the Aristotelian-based medieval synthesis of Thomas Aquinas, the thought of the church in the middle Renaissance was far from monolithic. Ficino both exemplified and promoted a resurgent Platonist inspired philosophy of religion. For Ficino, as for the neo-Platonists of various sorts, the philosophers of ancient Greece (and also, as far as we can gauge, for the even more ancient Babylonians) philosophy, cosmology and religion were all of a piece.

Ficino considered Pythagoras (c.582 - c.507 B.C.), Socrates and Plato to have been "religious philosophers". They, along with the prophets of the Old Testament, were precursors of Christianity. Like those prophets, the philosophers also, he assumed, would share in eternal salvation. Platonic tradition, therefore, had a proper place in the divine scheme of history.[150] Ficino, like the neo-Platonist Hermeticists Patrizzi and Bruno of later date, was convinced that the teachings of Plato were able to further religion and even bring people back to the Christian faith. In addition he believed that theology proper began in the pre-Platonic tradition of Hermes Trismegistus. As he writes in the "Preface" to his translation of the Hermetic documents, "Mercurius Trismegistus was the first philosopher to raise himself above physics and mathematics to the contemplation of the divine."[151] Then after stating that Trismegistus was considered the founder of theology he goes on to name Orpheus, Aglaophemus, Pythagoras and Philolaus as successors. He then states, "So six theologians in wonderful order, formed a unique and coherent succession in ancient theology, beginning with Mercurius and ending with the divine Plato."[152]

A basic aspect of Plato's theological philosophy that allowed Ficino both to feel at home with and to succumb to neo-Platonism and Hermeticism, was the ubiquitous doctrine of harmony and unity of all things. We have met this concept many times before: reality was thought to consist of a hierarchical but fluid structure wherein God, as the Good and author of the Good above, infused himself into the world below to participate in and form all things. God, who was responsible for himself and for all reality, was perfect in himself. Nevertheless, his being was reflected in the lesser beings on earth. As God moved earthward from his place of transcendence, he was met by the souls below who, being a part of the divine even under the exigencies of the material, reached above themselves and aspired to move away from earth back to the place of their origins. There they rejoined the divine and in the end were reabsorbed into divinity itself.

The inspiration for the scheme which, as we have seen, was repeated again and again in Plotinus, Proclus and the Hermetic writings, is found in Plato himself. The power behind creation is referred to as "God", "mind", or "the first principle". In the *Timaeus* they are synonyms. This is true whether this power is the good, the form of the world or the world itself.[153] In the *Laws*, God is the "ruler of the universe". He rules all things with a persistent desire for preservation and with a teleological view toward perfection. Plato's presiding ministers, his subordinate gods or intermediate powers, are responsible for every action.[154]

As we have seen repeatedly in Plotinus, Proclus and Hermeticism, these intermediate powers, intelligences, souls or gods interpenetrated and participated in all of nature. They are responsible for its movements, its generations and changes moving from the spheres of the heavenly bodies to the people and elements of earth. As souls they will eventually slough off their earthly dimensions and ascend to the Father from whom they came, to whom they will eventually return, and in whom they are to rest in eternal repose even as they reassume their pristine divinity.[155]

As for Whitehead in our time, so for Ficino in the sixteenth century, the ideas of Plato were of primary import.[156] Ficino deemed two aspects of Plato's theological thought especially valuable: "a pious worship of the known God and the divinity of Souls".[157] Taken together, Ficino suggested they constituted the understanding of things, all institutions of life, and all happiness.

Like a true Platonist and Hermeticist, Ficino conceives of God in terms of *being*. God is "the first being", the *primum*, the highest being, the *summum*, or "being itself".[158] He also speaks of God as the cause of being for all things.[159] Yet, whereas Plato understood that the world had its source with the divine being, Ficino Christianized Platonism to the extent that in creation God produced "something out of Nothing into being".[160] God himself "is the absolute Being . . . itself, in other words pure act, cause of all existing things".[161] God is thus "the Being above all, that Being which is conceived with some proper concept of Being and understood by a notion".[162]

Ficino sees correctly that for the Platonist (as well as for the neo-Platonist and Hermeticists, he might have added) there are real gradations in God but no differentiation. The highest god is unity or goodness. The next god is being. Thus Ficino was convinced that all gradations, perfect and imperfect, good and evil, the highest and the lowest depend upon their relationship to being. "Such gradation in existence does not happen, nor can it be known except by proximity to the highest Being, which is God Himself, and likewise by distance from it."[163] God's existence, therefore, is not determined "by a certain species of existence, through which it would become a particular being.[164] Nevertheless, "Being itself is common to all things. Wherever there is Being it depends on God".[165]

> In all genera of things there is one greatest and highest and by

participating in it the other things are placed in the same genius: as, for instance, all warm things become warm through the nature of fire, to which the greatest warmth is intrinsic, and all good things must be called good because they follow and imitate the highest and first good.[166]

Again, the world receives its order from God himself. It is ordered as it emulates him. God determined by his own will to create the things of the world. Therefore, in essence, all share in his goodness and perfection from the beginning.[167] In a way that is unexplained, the world consists of "lower things." God is active in the world moving it toward perfection.[168] Both the purpose of things and their activities are therefore directly related to the divine. "It is the task of the same thing to perfect and to produce."[169] God, of course, as the constant creator produces and perfects all things.[170] "God creates by degrees things more or less similar [to himself]" so that all things are "fulfilled by the highest reason" which is God himself.[171]

Ficino's scheme is essentially *teleological*. Perfection results from the process as it reaches toward its proper goal. Although in both Platonic and neo-Platonic conceptualities, perfection requires purity, [172] for Ficino the emphasis upon the *essential harmony* between God and that which he has created is such that the order of nature itself is basically good. There is that, however, which is contrary to the order of nature. This is defined as evil.[173] However, because the will is directed by its nature toward the good,[174] evil is not a substance but rather a limitation. It is an insufficient participation in the good. It can, for example, be described as "blindness".[175]

Thus Nature is all-sufficient. The statement: "Nature is not lacking in necessary things, nor is it abundant in superfluous ones" (*"In necessariis rebus natura non deficit, supervacuis non abundat"*)[176] occurs almost word for word in Aristotle's *De Anima*. Here Aristotle asserts, "nature does nothing in vain, nor omits anything essential.[177] Whereas in Aristotle the statement refers to the proper disposition of an organism of nature for its own perfection, in Ficino the assertion is applicable to the whole of nature as such. According to

Ficino *goodness*, at one and the same time, is inherent to nature in that it is a response to *desire*. Desire, in turn, is a part of nature. Like nature, it comes from God and shares in the good.[178]

Ficino's harmonious universe which is good in itself but which at the same time moves toward goodness, is almost a carbon copy of the universe of Plotinus, Proclus, and the Hermeticists. In these schemes the perfection of a thing comes about when its own qualities are properly integrated in the movement of each thing toward its origin. Thus its end and beginning are identical. The universe is in constant movement and each movement "leads to a good".[179] As natural appetites tend toward satisfaction, so all creation tends to the good.[180] "Rational Souls" that, for Ficino, participate in God, are "intellectual by participation", even as they are "angels by form" and "Ideas finally by cause".[181] Rationality is analogous to light that not only transports "the forces of the stars to . . . things, but it brings the sun and the stars themselves to lower beings".[182]

> In a similar way the Soul of man is sent from God to matter, penetrates it in one moment, but does not leave God. . . directing the body, and, through intellect, attaining the truth of all things, namely, God . . . Again, as light is reflected into the sun, so the Soul is reflected to God through its will, desiring always the goodness of all things, God.[183]

In addition Ficino can say:

> The leader of nature gave to the Soul the desire for the universal and whole true and good, which is more natural than the desire for food and copulation, since it is more continual. For the body seldom requires food, and still more seldom copulation. But we desire the true and the good every single moment.[184]

Hence, as "plants leave and regain their natural habitations" and particles return to the place from which they have separated, souls themselves eventually return to their natural states.[185] Final destiny is assured.

> [God] would be too rash and inexperienced a marksman if He directed our desires toward Himself like arrows toward a mark and had not added feathers to the arrows by means of which they might some time attain the mark. He would be unfortunate if His attempt through

which He attracts us toward Himself never reached its end.[186]

This teleological movement is exemplified *par excellence* by the movement of the soul. The soul is like the fire, moving upward "because the higher place is good for the fire, the fire is therefore moved toward it and rests in it".[187] Hence the soul's desire is to return to God and in the end to become God. The "ultimate end of man consists in the knowledge of possession of God alone which alone ends his natural appetite".[188] However, even as in Platonism and its derivatives, so in Ficino the scheme is not perfect. There is the abyss or Hades for souls that are evil.

> All [virtuous Souls] are moved by a corresponding habit, like a natural lightness, toward that region which is the dwelling place of the angels after whom they have most closely patterned themselves during life. In like manner Christians believe that by their similarity, like a natural weight, damned Souls dash headlong toward the nine degrees of damned demons to which they made themselves similar during life.[189]

A state of damnation is thus possible. It is nevertheless the non-normal and unnatural state of the soul. For the soul, like the double-faced Janus, looks at both the corporeal and the incorporeal.[190] It is "on the border-line between eternity and time" turning both "toward eternal and toward temporal things"[191]. The soul, therefore, is the natural form of the body.[192] Even though it will eventually rest in God and even after it has departed from the body, "the Soul is left with a desire to return again to the same body".[193] "Souls remain eternal even after the destruction of the body at some time . . . they will again receive their bodies".[194] In the end, therefore:

> The body will be resurrected to become entirely immortal. For from the beginning God disposed the order of things in such a way . . . that through the Soul always living and always naturally desiring to animate, the body likewise may always live.[195]

The resurrection of the body in Ficino as in Christian theology coincides with the end of time.

> This creation of God will be directed to something stable when the movement of the world introduced for the sake of some more perfect

rest, ceases. Therefore the body will always remain united with the Soul.[196]

Ficino's writings provide ample evidence for Kristeller's argument that Ficino believed that he was an instrument of divine providence just like Plato.[197] Consequently it is hardly a coincidence that he considered it a part of his mission to enter into a crusade against Aristotelianism. Hence Ficino complains:

> The whole world is now in the hands of the Peripatetics and is divided mainly into two sects: Alexandrists and Averroists. Both deny any form of religion. If anyone thinks to destroy by the simple preaching of faith an impiety so diffused among men and defended by such subtle minds, he will soon be refuted by the results. Stronger measures are needed: either divine miracles manifested on all sides or at least a philosophical religion to which philosophers will listen more readily and which will some day succeed in convincing them. But in these times it pleases divine Providence to confirm religion in general by philosophical authority and reason until, on a day already predestined, it will confirm the true religion, as in other times, by miracles wrought among all peoples.[198]

Whether in the end Ficino was more influenced by Platonism, neo-Platonism or Hermeticism is open to question. There would seem little doubt, that the powerful current of the Platonic tradition that combined mystical elements from Hermeticism with mysticism was integrated into his thought with elements of biblical teaching. In Ficino's era and through Ficino himself, this mystic para-rationality was streaming into the Renaissance mind and calling for a new synthesis with the Christian Faith.

Although the earlier tradition of Christian Platonism founded its thought on the legacy of Platonic ontology, Ficino is the first to attempt to integrate the totality of this tradition with a Christian understanding of reality. Attractive as this attempt was it eventually failed due to the fundamental epistemological incompatibility of the conceptual complexes involved. It did so in a way similar to Thomism which sought to mix the water of Christian doctrine with the oil of Aristotelian dogma. Ficino's attempted synthesis offered a degree of freedom for speculation and it inspired a desire to penetrate the unknown mysteries of the universe. In addition to his attempt to understand and to worship the God of the Christian faith, Ficino encouraged the Renaissance mind to think beyond the rationalism of Aristotle into the yet unexplored and unknown mysteries of the universe.

With regard to cosmology in particular, one is tempted to speculate as to the degree Ficino was influenced directly or indirectly by Hermeticism. His writings are based largely on Plato and Plotinus. He repeats again and again such common Platonic notions as the interpenetration of God in all aspects of the universe, the participation in all things in the divine, and the harmonious interaction of all things. This harmony is not dependent upon outside forces, but is due to the conception that objects were possessed of self-animation and self-direction. Also the idea that all things and persons reach their final destiny of being reunited with Being itself played a central role in Hermeticism. To the Renaissance mind, however, Hermes has the definite advantage over Plato because it was believed that he was more ancient. Ficino, like thinkers from the third century onward, regarded him as the very foundation of Plato, the source of the kind of thinking that Plato brought to perfection.

Hence, Ficino has no difficulty referring to the *Hermetica*, and the writings of Plotinus and Plato on the subject of higher and lesser goods within a single genus. For him, the authority of ancient thought was found to be equal to that of Scripture.[199] He understands cosmology in terms of Plato, the Hermetic writings, and Psalm 19.

> The celestial entities tell you the glory of God through the rays of the stars, like the glances and signs of their eyes, and the firmament announces the work of His hands.[200]

It is in regard to the sun that the Hermetic influence shines through.

The sun can signify to you God Himself in the greatest degree. . . . So the invisible things of God, that is the angelic divinities, are seen and understood particularly through the stars, and *God's eternal power and divinity through the sun.*[201]

The Hermeticists didn't invent the heliocentric theory, of course. It was first put forward, as far as we know, by Aristarchus of Samos in the third century B.C.. The fact that the theory is mentioned by Plutarch (c.45 - c. 120) would indicate that it had been known in the West since the time of the Church Fathers. The concept of heliocentricity was, in all likelihood, generally known, although not accepted, from Aristarchus onward. Thus, it is not surprising that Nicholas Copernicus (1473 - 1543) mentions Aristarchus in the first manuscript of his *De Revolutionibus* although the reference is crossed out in the copy sent to the printers.[202]

Ficino's book *Liber de Sole* (On the Sun) was printed in 1493. During the previous year he had defended a theory of heliocentric cosmology at the University of Bologna in a debate against the Averroist Alexander Achillini (1463-1512) who championed geocentric cosmology. In 1497, just four years after this renowned debate, Copernicus took up residence at the University of Bologna to pursue his astronomical studies.[203] We can judge the content of the debate between Ficino and Achillini by the latter's *De Orbibus (The Orbits of the Planets)*, published in 1498. *De Orbibus* was a defense of geocentricity as based on Averroës' [Ibn Rushd] (1126 - 1198) commentary on Aristotle's *De Caelo*. In what must be regarded as a reply to Ficino, Achillini argues resolutely for geocentric cosmology and makes it quite clear that from his standpoint it is absolutely illegitimate to place the sun at the center of the world.[204]

Ficino himself argued that the sun is the "significant" God, placed at the heart of the world. The sun is the pilot that guided the heavenly bodies, the measure of their divinity, and the body responsible for nourishing and generating life on earth.[205] This argument is `identical to the heliodominated and heliocentric universe proclaimed by Trismegistus, the Egyptian Moses, in the *Hermetica*. In his *De*

Revolutionibus (1543) Copernicus follows his heliocentric diagram of the universe with an ode to the sun. The ode not only mentions Trismegistus but uses Hermetic terminology in its description.

> In the center of all rests the sun. For who would place this lamp of a very beautiful temple in another or better place than this from where it can illuminate everything at the same time? As a matter of fact, not unhappily do some call it the lantern; others, the mind and still others, the pilot of the world. Trismegistus calls it a "visible god"; Sophocles' Electra, "that which gazes upon all things." And so the sun, as if resting on a kingly throne, governs the family of stars which wheel around.[206]

Such lines, poetic though they may be or perhaps exactly because they are poetic, should convince us that Copernicus, as well as Ficino, was familiar with the Hermetic sun-centered world. Thus Copernicus was not the first to conceive of the sun in the middle of the world. Rather, his genius consisted in giving the heliocentric system a plausible (barely plausible) geometric argument. Thus, he substantiated by way of mathematics an idea that was considered to be "heliocentric speculation" until the time of Galileo and Kepler some seventy years later.[207]

To return to the theme that we introduced at the beginning of this chapter, the Platonic, neo-Platonic and Hermetic doctrines offered an alternative understanding of the world to that of the deductive system of Aristotle. It was on this foundation that the "New Science", to use A.-J. Festugière's term, was to be built.[208]

The "New Science"
Given the mystical, non-differentiable, interwoven and interconnected concept of nature presented by the Platonic, neo-Platonic and Hermetic conceptions of the world, it would appear improbable that these esoteric forays into the realm of the fantastic could contribute to the development of "modern science". Yet the evidence is that it did. This pattern has a precedent. In the ancient world, because the stars were thought to reflect the ways of God, the

Babylonian and Pythagorean interest in the heavens was primarily ritualistic. Eventually the canons of worship which centered on the heavens, paved the path to astronomy. Similarly, in the Renaissance the imaginary and mystical philosophies of the Platonic, neo-Platonic, Pythagorean, and Hermetic traditions, characterized by linkings and interconnections between the real and eternal aspects of the world and its less-than-real transitional material aspects, served to open the Renaissance mind to the arena of scientific investigation.

One of Platonism's contributions to the understanding of the world as we now know it was a high regard for Pythagorean mathematics. It will be recalled that neither Plato nor the early Pythagoreans used numbers to quantify the phenomena of the material world. For the Pythagoreans, Plato, and interestingly enough, for at least one branch of Arab thought, that of the Brethren of Purity, numbers were used only secondarily for counting. Primarily, numbers were understood as comprising the essence of reality. As the heavens were thought to reflect being, so, too, being was understood as numerical. As Plato records in the *Timaeus*, numbers designated the intervals between the heavenly spheres according to two geometrical progressions: 1..2..4..8 and 1..3..9..27.

In each interval there were two kinds of means:

> the one exceeding and exceeded by equal parts of the respective extremes...the other being that kind of mean which exceeds and is exceeded by an equal number.[209]

In this way the heavenly harmonies not only reflected the circular movements of the heavenly bodies, they also showed forth their symmetry in the geometric ratios of the spaces between them.[210] Further, the numbers that were exhibited by the heavens, like the circular shape of the celestial movements, were of the very essence of Plato's world-soul and the basis of order and rationality. In assigning the earthly elements to geometrical figures — the earth to the cube, the icosahedron to water, the octahedron to air and the tetrahedron to fire — Plato used numbers to bridge heaven and earth. He then assigned reason over both spheres.[211]

> When the creator had framed the soul according to his will, he formed within the mind the corporeal universe, and brought them together, and united them center to center. The soul, interfused everywhere from the center to the circumference of heaven, of which she is the external envelopment, herself turning in herself, began a divine beginning of never-ceasing rational life enduring throughout all time.[212]

In that the visible heavens show forth the reason and harmony of the soul, to know the heavens by number was to have insight into the perfection and harmony of eternity and of divinity itself.[213]

Thus, for Plato, although numbers and mathematics could be used for the calculation of earthly things, it was *the soul* that made proper use of numbers. Numbers were the "easiest way to pass from generation [*i.e.*, the things of earth] to truth and being."[214] Arithmetic, therefore, is "truly necessary" as a subject of study, because it requires the use of pure intelligence to attain pure truth.[215] In the same way, Plato considers that geometry too is useful for pitching tents and the like, but its practical application is quite secondary. "The knowledge at which geometry aims is of the eternal and it is not of the perishing and the transient."[216] Geometry will, in fact "draw the soul toward the truth".[217] *Mutatis mutandis* the primary purpose of astronomy is not to know the composition of the heavens. Rather, "astronomy compels us to look upward, and leads us from this world to another".[218] It leads us to contemplate the divine. "The starry heavens are to be used as a pattern of that higher knowledge".[219]

In reflecting the Pythagorean belief that the different heavenly bodies issued forth different melodies as they moved in their eternal circles, Plato was convinced that even as "the eyes are appointed to look at the stars, so are the ears to hear harmonious motions".[220] He knew, of course, that there was a danger that the "sciences" would take on a practical aspect and, while looking toward the heavens, would make the mistake of contemplating the earth. He was however convinced that such misdirected efforts entailing as they did learning something from the senses, would cause the soul to look not upward but downward.[221] *Proper science*, to the contrary, is not that which uses the sight but that which uses the mind only. The mind teaches "the eye of the soul . . . to look upwards".[222]

As we have seen, such themes are reiterated again and again in Hermetic thought. They are also basic to the *Rasa'il* (*The Summary of Knowledge*) produced by the Arab mystical cult, The Brethren of Purity, whose writings were a part of the legacy the Arabs bequeathed to the West. In contrast to most Arab "scientific" thought, based as it was on Aristotle, the thought of the Brethren of Purity was influenced by neo-Pythagorean, neo-Platonic and Hermetic philosophy.[223] The *Rasa'il* itself is an expanded emulation of what we know of the mathematical concepts of the Pythagoreans. Mathematics and number were characterized as being the essence of reality. Number, geometry, astronomy, geography, and music were all considered to be sub-units of mathematics. Numbers assigned to the different Arabic letters functioned as a secret code which when decifered revealed the symbolic significance of the text of the Koran. Spiritual powers, heavenly spheres and earthly elements were all considered to reflect different numbers. Up and down the great chain of being from the Creator to the creature and including the world of plants, animals and minerals, numbers designated the levels of the hierarchical reality. They designated the relationship of each level to every other, and connected all to the Creator.[224]

Following the intellectual traditions of neo-Platonism, of Hermeticism and to a certain extent Aristotle, The Brethren of Purity were convinced that the heavenly powers controlled the destiny of the earth. They believed that all things interpenetrate one another and change into one another.[225] This conviction that was basic to Plato's understanding of the material world, had a direct effect upon Arab alchemy and upon the alchemy of the West. The popularity of this *black art* in the early Renaissance can be judged by the fact that Robert of Chester's (fl. 12th cent.) translation of the

Arabic *The Book of the Composition of Alchemy* (1144) was but the first of a half dozen other books on alchemy to be translated.[226] Like Plato who considered the elements which constitute the world — fire, water, air, and earth — to be impure and not possessing the power worthy of their nature and dependent upon the "universal fire",[227] so the alchemists attempted to use fire to perfect the elements or to change one element into another. So convincing was the fundamental argument which lead to alchemistic attempts to change metals into one another and especially to metamorphose sulphur or mercury into gold that people from Roger Bacon in the twelfth century to Francis Bacon in the seventeenth and even Isaac Newton in the eighteenth were practitioners of the "black art".[228]

The word *alchemy*, as we have seen, goes back to the Arabic *al kimiya* with the article, *al*, as a prefix to *Khem*, which may either refer to the Greek Αἴγυπτος for Egypt or the Egyptian *chemi* which means *black*. Hence the "black art", and later the word *chemistry*.[229] The actual technique involved in alchemy may have evolved from early Egyptian metallurgy. This skill was first distinguished for its fine golden jewelry and later for attempts at imitating precious metals with amalgams of lesser value. A method of verification developed which involved heating the metals so that an actual alteration of the color of the metal took place, thereby giving it a superficial patina. The technique that evolved for the processes used in falsification as well as those used in detecting fake metals by the use of heat led to the actual change of the color of the metal. Eventually the technique also led to the plating of silver with gold and the fusion of two metals, such as gold and silver, into an alloy which looked like gold, but was harder and more durable than pure gold.[230]

According to the philosophy of alchemy, all elements are essentially of one Being and flow in and out of one another in nature. Consequently, practical alchemy became an attempt to surpass nature. It sought to mimic what nature does under the power of the sun, but at a faster rate and by the application of earthly fire. Alchemy was thus understood as a process that cooperated with nature. A further step was based on the Platonic idea, put forward by Aristotle as well as by the neo-Platonists and the Hermeticists, that earthly elements not only contain the divine essence but they also are essentially of divine substance but in material form.

The consequences were clear. Since material was but the shadow of the divine, it should be a simple matter to use fire to distill from the elements the divine essence, the *quintessence* or the *Philosopher's Stone*. The process promised to provide the *divine tincture* by means of which base metals would be brought to perfection and so become the perfect metal, gold. By the same token, ailing bodies could be restored to health and even souls could be cured and saved.

Thus the experimental technique of alchemy may be the first technique of what Festugière called "the new science" that arose in the late Renaissance. Although the "new science" rested upon trust in revelation, the fact that God was manifested not only in the heavens but also in the whole of earthly reality led to the belief that it was possible to discover the divine within the material. Hence, whereas the "old Aristotelian science" was disinterested in nature and had little practical application or promise of immediate benefit for the individual, the "new science" was supremely interesting and self-involving. It supported intelligent investigation and attached profit to the knowledge it promised to deliver. Astrology offered a vision of human destiny. Alchemy promised the production of gold, of becoming fabulously wealthy, just as it held out hope for the health of the body, and the salvation of the soul. Even the practice of magic was a serious attempt to control the destiny of the world and of humankind.[231]

In contrast to Aristotle who argued deductively from the general to the particular, the "new science" brought about a new logical orientation. Because all particulars flowed from and contained the essence of being themselves, to know the *particular* thoroughly was to know not only the *one* but to know the *many* and the *all* as well. Hence concentration was placed upon knowing the individual object under the supposition that the truth of any particular would be truth in general. The same divine essence that is in *everything* is the essence of *all things*.

Yet the "new science" was not totally anti-Aristotelian any more than Aristotle was totally anti-Platonic. Thus, as in Aristotle, for whom *causes* were known and the *material* was simply a manifestation of cause, an *accident* of *real substance*, so too in the "new science" material objects were shaped, formed, and accompanied by the divine substance within them. In the "new science", in contradistinction to Aristotle, materiality too was essentially good. *Things* were a combination, almost an amalgam, of the divine and the material. Thus, rather than material things being considered purely *accidental* they were considered important and not to be despised.

At the same time, the presence of the divine souls, intelligences or forms in all things made nature itself redolent with mystery. Nature was a moving, undulating, cyclical affair, arranged in a hierarchy of being. It was a part of earth but was linked to heaven. In essence nature was divine. The lines of sympathy that led heavenward and the antipathy that caused it to fall back again to earth, were reflected throughout the whole.

Nevertheless a metaphysical inconsistency in the scheme allowed for a differentiation between the good and the bad, for, were everything fully divine, there would be no evil. Even so the conviction that the good would triumph impelled the search for the beneficial aspects of nature and the avoidance of that which was harmful. Metals were arranged into a hierarchy of the most perfect to the least perfect and plants were arranged according to the medicinal helpfulness. Astrology also.played a role, for the stars themselves were divine and controlled the affairs of humankind and nations.

One turned one's eyes to the heavens to understand the essential harmony of God and of all things as well as to divine one's own particular destiny. The contemplation of the heavens would not only "raise the soul", but knowledge of the paths of the stars and the planets enabled one to know when to plant, to harvest, to take medicine, to copulate, to abstain, and how to manipulate nature for desired effects. The sun, the primary representative of God in the heavens, the central divine body of the universe, bathed all things with its beneficial light even as it represented the central power of the creator who held all things in his power and blessed them with all that was good. Therefore, rather than being simply teleological as was Aristotelian science, the "new science" promised immediate and ongoing results. It was, as Francis Bacon was to say much later, to be practiced for the benefit of humankind.

However, because the "new science" dealt with a mysterious universe that was not only hidden from reason but which was both a part of and directly linked to the divine, the final appeal was to direct revelation. In that the "new science" was, in fact, a mystery, and was the transmission of mystery, like the "science" of the ancient Pythagoreans, it was linked to cultic practices and prayer. Little wonder that, like Pythagoras himself, the practitioners of the "new science", Ficino and Bruno and to some extent even Kepler, saw themselves as the priests of the new era, an era to which the new science promised to give birth.[232]

Thus, in the mystery-laden cultic neo-Platonic and Hermetically induced atmosphere of the Renaissance, "the new science" which dealt with the material world did not become a rigorous science at all. Because of the metaphysical inconsistency mentioned earlier, it was unable to investigate the material world with the type of resolute penetration of thought necessary for modern science. The material world was a mysterious world penetrated and given form by the essence of being, God. The divine world rather than the material world continued to be the *raison d'être* of the exercise. While it is true that the process of discovery necessary to the "new science" engaged the whole person, the fact that the person was immersed in the mystery of being of which reality consisted meant that the person was initially concerned with his own personal destiny. Discovery as such was of secondary interest. In the end the mystery won out. Observation was used in the Platonic and Hermetic traditions. Yet one lifted one's eyes to contemplate the heavens to meditate upon their divine secrets, not to count the stars. The insistence on a contemplative and cultic life was not directed by an authentic desire to learn about the material world in itself but by the desire to lift everything to the divine so that it would itself become divine.

To us, the whole process may well have had to do only with a world formed by the mystically-oriented imagination. Yet there is little doubt that, at the time, the "new science" served to open the imagination to new understandings of the world. It encouraged the inquisitive imagination to look beyond the phenomenological world and the world of classical Aristotelian science. In addition because God was constantly interpenetrating the material world with his goodness, the whole of reality could be looked upon in a positive rather than a negative sense. To paraphrase Romans 8:28, "all things work for good for those souls whose concentration was directed to the good". This gave those who were so directed a buoyant confidence that the One who penetrated reality with his divine goodness would eventually call all things to himself.

It was a faith in the God who manifested himself in the world especially in the sun. For even as the sun in the center of the universe controlled the spheres of the planets through its light and caused the generation of all things on earth, God who is light infused himself into the lowest depths of the world to generate the good in all things and to call all souls back to himself. With him and in him, who is both their origin and their final destiny, these souls would rest in perfect peace and await the resurrection of their bodies with which, at the end, they would be eternally re-incorporated.

The "new science" was brilliant. It was beneficial to the development of modern science and an inspiration to some of its pioneers. Nonetheless its mystical understanding of reality could not and did not give rise to experimental science. Unlike the inconsequent and "holistic" methodology of the "new science" that is grounded in the contemplative life and inspired by the occultism of alchemy, experimental science requires differentiation, quantification and the use of both inductive and deductive reason. Nevertheless, the "new science" along with the philosophies on which it was based did inflame the Renaissance mind not only with the insight of the importance of the central fire of the sun but it aroused the imagination as well. The understanding of nature as a harmony that could be probed for its mysteries inspired the imagination to think beyond the *known* to the *unknown*. Eventually it became a positive factor in the rise of modern science.

Before that could happen, however, it was necessary that its wild, uncontrollable and uncontrolled flights of imagination be tamed and brought back to earth. Its insights needed to be threshed and winnowed by the teachings of the Christian faith as understood and propagated during the Protestant Reformation of the sixteenth century.

NOTES

[1] A. C. Crombie, *Augustine to Galileo*, 2 vols. in 1 (London: Heinemann, 1979), I, 51 f.
[2] John Marsh, *The Fullness of Time* (London: Nisbit, 1952), p. 17.
[3] Thorlief Boman, *Hebrew Thought Compared with Greek*, (London: SCM Press, 1960), p. 18.
[4] Plato, *Timaeus, Dialogues of Plato*, v. 2, 36-40; Aristotle, *On the Heavens*, I.ix, 279a; II.i, 284a; II.iv, 287a; II.xii, 292a-292b.
[5] Aristotle, *On the Heavens*, III.iv-viii, 302b-307b.
[6] Aristotle, *Parts of Animals*, (LCL, 1937), I.i, 640b.
[7] His "quarrel" with such physicists as Anaximenes, Anaxagoras, and Democritus was that when they spoke of the motion of the earth, for instance, they concerned themselves with "the particular parts" rather than with "the undivided whole", Aristotle, *On the Heavens*, II.xiii, 294b.
[8] Aristotle, *Parts of Animals*, I.i, 641b.
[9] Aristotle, *On the Heavens*, II.x, 291b.
[10] Cf. Stanley L. Jaki, *The Relevance of Physics*, pp. 14 ff.
[11] E. T. Whittaker, *From Euclid to Eddington* (Cambridge: Cambridge University Press, 1949), p. 46.
[12] For a brief but lucid explanation of the axiomatic method, cf. J. R. Carnes, *Axiomatics and Dogmatics* (New York: Oxford University Press, 1982), pp. 19-37.
[13] Cf. Nebelsick, *Circles of God*, pp. 65, 70.
[14] In 1895 H. Koch and J. Stiglmayr recognized that the Dionysian doctrine of evil was strictly neo-Platonic and they therefore concluded that these writings could not have been written prior to the fifth

century. More telling, perhaps, was the evidence that the first indisputable citation from the works was made by Severus of Antioch between 518-528. Cf. the *New Catholic Encyclopedia* on Pseudo-Dionysius.

15 Raymond Klibansky, *The Continuity of the Platonic Tradition* (London: Warburg Institute, 1950), p. 25.
16 *Ibid.*
17 Cf. Nebelsick, *Circles of God*, pp. 155-158.
18 Dante Alighieri, *The Banquet (Il Convito)*, tr. Katharine Hillard (London, Kegan Paul, 1889), II. VI,4, p. 78. Cf. Nebelsick, *Circles of God*, pp. 158-162.
19 Marsilio Ficino, *De Christiana Religione*, Ch. XIV cited by Frances Yates, *Giordano Bruno*, pp. 118 f.
20 Cf. F. E. Peters, *The Harvest of Hellenism*, (New York: Simon and Schuster, 1970), pp. 350 ff.
21 Cf. above, p. 77.
22 Francis Bacon, *Novum Organon, The Works of Francis Bacon*, 7 vols. (London: Longman & Co., 1857-1858), Book I, Aphorism LXXVII.
23 Nebelsick, *Circles of God*, pp. 4-9.
24 A. E. Taylor, *Platonism and Its Influence*, p. 4.
25 Cf. below, p.
26 Plato, *Timaeus*, 68-69. For a sympathetic rendering of Plato in relationship to the Christian faith, one, however, that continues to stress the essential difference between the Judeo-Christian doctrine of creation and that of Greek philosophy in general including Plato, cf. Diogenes Allen, *Philosophy for the Study of Theology*, especially pp. 15-59.
27 *Ibid.*, 34, 68.
28 *Ibid.*, 36.
29 *Ibid.*, 68.
30 Plato, *Parmenides, Dialogues of Plato*, vol. 3, 142.
31 *Ibid.*, 144-145.
32 *Ibid.*, 145.
33 *Ibid.*, 147-148.
34 Nicholas of Cusa, *De docta ignorantia (Die belehrte Unwissenheit)*, 3 vols. (Hamburg: Felix Meiner, 1970-1977), I. Cap. XXIII.71.
35 Plato, *Parmenides*, 145. Cf. Cusa, *De docta*, I, Caps. XVI, XVII and XII.162.
36 Plato, *Timaeus*, 28-56.
37 Plato, *Laws, Dialogues of Plato*, Vol. 4, X. 903.
38 Plato, *Timaeus*, 39-41; Cf. Aristotle, *Metaphysics*, X-XIV, XII. viii-ix, 1074b.
39 Cf. Nebelsick, *Theology and Science*, p. 53; and *Religion in Geschichte und Gegenwart*, 3rd ed., 7 vols. (Tübingen: C. B. Mohr, 1961), V, 36.
40 Nebelsick, *Theology and Science in Mutual Modification*, "Process Thought", pp. 45-62.
41 Plato, *Laws*, X. 889.

140   The Renaissance, the Reformation and the Rise of Science

42 *Ibid.*, X. 908-909.
43 *Ibid.*, X. 904.
44 Plato, *Philebus, The Dialogues of Plato*, Vol. 3, 55.
45 Plato, *Republic, The Dialogues of Plato*, Vo. 2, 527.
46 Plato, *Philebus*, 61.
47 Plato, *Philebus*, 61.
48 Plato, *Timaeus*, 28 ff. Cf. Nebelsick, *Circles of God*, "Platonic Ethereality", pp. 18-23.
49 Plato, *Timaeus*, 27-30, 44-47.
50 Plato, *Theaetetus, The Dialogues of Plato*, Vol. 3, 186-187, cf. 150-161, 179-182, 185.
51 *Ibid.*, 186.
52 *Ibid.*, 187. It is true, of course, that scientific knowledge only arises by intuitive insight into the data of perception rather than directly from the data themselves, but without the data of perception, there can be no knowledge at all.
53 *Ibid.*, 186.
54 Cf. Nebelsick, *Circles of God*, "Pythagorean Cultic Cosmology", pp. 9-17.
55 Cf. Max Planck, *Where is Science Going?*, pp. 214 ff. Planck's belief in causation continues to identify him with the Newtonian world-view, but his description of the process of discovery is worthwhile indeed.
56 Cf. Nebelsick, *Circles of God*, "Platonic Ethereality", pp. 9-17.
57 Plotinus, *Enneads*, 2 vols, tr. Stephen Mackenna (Boston: Branford, 1917), I.6,1; I.8,5; III.5,5; IV.1,1; IV.3,27-29; IV.7,2; IV.8,6; IV.9,1; V.4,1; VI.1,1-VI.3,28; VI.7,38; VI.9,9.
58 *Ibid.*, I.1,1-I.7,13; II.2,1-3; II.3,1-18; II.5,2-5; II.6,1-3; III.7,4-5; V.2,1.
59 *Ibid.*, IV.7,2; IV.8,6; V.1,10.
60 *Ibid.*, Vol. I, "Porphyry's Life of Plotinus", par. 7, p. 9.
61 *Ibid.*, par. 16, p. 15.
62 *Ibid.*, par. 20, p. 19.
63 *Ibid.*, par. 14, p. 13.
64 *Ibid.*
65 *Ibid.*
66 *Ibid.*, par. 10, p. 11. Italics mine.
67 Plotinus, *Enneads*, VI.9,11. Cf., *ibid.*, III.7,11; III.2,5-7; VI.1,1-VI.3,27; VI.8,3-9.
68 *Ibid.*, II.9,1; Cf., *ibid.*, IV.9,1-5; V.1,10.
69 *Ibid.*, V.1,5; VI.9,9; VI.9,10.
70 Cf. Paul Tillich, *Systematic Theology*, 3 vols. (Chicago, University of Chicago Press, 1951-1963), vol. I, pp. 205, 235 and vol. II, p. 293 for "ground of being". Cf. *ibid.*, vol. I, p. 289 and vol. II, p. 294 and Paul Tillich, *Biblical Religion and the Search for Ultimate Being* (Chicago: University of Chicago, 1955), p. 82 for "being itself". Cf. Paul Tillich, *My Search for Absolutes* (New York: Simon & Schuster, 1967), pp. 82, 127 for "power of being". Cf. also, p. 137 for "the absolute itself", and *Systematic Theology*, vol. III, p. 410 for "essentialization".

The Renaissance Mind

71 Plotinus, *Enneads*, V.1,1; VI.9,6; VI.9,11.
72 Augustine, *The Confessions of St. Augustine*, tr. F. J. Sheed (New York: Sheed and Ward, 1943), I. 1, p. 31.
73 Plotinus, *Enneads*, I.7,2; VI.9,11.
74 The reference is apparently to the Gnostics.
75 *Ibid.*, II.9,1; II.9,16.
76 *Ibid.*, I.7,3; II.1,2; II.1,3; III. 8,2.
77 *Ibid.*, II.1,1.
78 *Ibid.*, II.1,2.
79 *Ibid.*, II.1,3; II.2,2; II.2,3.
80 *Ibid.*, IV.7,14; IV.8,3.
81 *Ibid.*, V.1,2; V.5,4; VI.9,1.
82 *Ibid.*, VI.9,3.
83 *Ibid.*, VI.9,6. Cf. Tillich, *My Search for Absolutes*, pp. 127, 137.
84 Plotinus, *Enneads*, VI.9,11.
85 *Ibid.*, II.2,3.
86 Cusa, *De docta Ignorantia* II, Cap. XII. 163; Plotinus, *Enneads*, II.3,3. Cf. Nebelsick, *Circles of God*, p. 189, fn. 209.
87 Plotinus, *Enneads*, II.3,7.
88 *Ibid.*, II.3,13.
89 *Ibid.*, VI.9,11.
90 Proclus, *Ten Doubts Concerning Providence*, tr. T. Taylor (London: Taylor, 1833), p. 1
91 *Ibid.*, p. 3.
92 Cf. Proclus, *The Elements of Theology*, tr. E. R. Dodds (Oxford: Clarendon Press, 1963), "Introduction", pp. xi-xiii. Interestingly enough, in the late Renaissance, it was the neo-Platonically-inspired texts of Giordano Bruno that were outlawed.
93 Proclus, *Ten Doubts*, pp. 4 ff. Plato, *Phaedrus, Dialogues of Plato*, Vol. 1, 246.
94 Proclus, *Elements*, 100-110.
95 *Ibid.*, 20.
96 Proclus, *Commentaire sur le Timée*, 5 vols., tr. A. J. Festugière (Paris: Vrin, 1966-1967), I, Bk. I, 53.8-54.10; III, Bk. 3, 102-3-316.4; V, Bk. V, 242.9-271.27
97 Cf. Tillich, *Systematic Theology*, II, pp. 9, 10, 87, *et al.*;III, pp. 99, 142, 190, *et al.* Cf. E. R. Dodd's, "Introduction", *Elements*, pp. ix-xiv for the character and purpose of *The Elements of Theology*.
98 Proclus, *Elements*, 1-6. Cf. Dodd, *ibid*, Commentary on Propositions 1-6, pp. 187 ff.
99 *Ibid.*, 7-13.
100 *Ibid.*, 14-19. As E. R. Dodds, translator of *The Elements of Theology*, points out, the neo-Platonic "trinity of subordination—Soul, Intelligence, and the One—is a combination of Platonic and Aristotelian ideas. The *Soul* is that of the world-soul of Plato's *Timaeus* and *Laws*. The *Intelligence*, like Plato's *secondary causes* in the *Timaeus*, is responsible for the creation of the world of things. Plotinus had identified

the *Intelligence* with Aristotle's *Nous*. The *One* is identical to the "One" in Plato's *Parmenides*. There, as in Proclus, it is the form of the Good. Dodds, Commentary on Prop. 20, *Elements*, p. 206. Cf. Plato, *Timaeus*, 30, 34, 68-69; *Laws*, X; *Parmenides*, 137 f.

101 Proclus, *Elements*, 20-46.
102 *Ibid.*, 47-69.
103 *Ibid.*, 70-74. It would seem rather obvious that Paul Tillich must have read Proclus although he has no reference to him throughout the three volumes of his *Systematic Theology*. He does, however, refer to Plotinus on three occasions in Vol. I, pp. 72, 141, 233, and once in Vol. II, p. 140 but only in a general way. Tillich's ideas on Being, participation in Being, ground of Being, would seem to be so close to those of Plotinus and Proclus that pure coincidence here is highly improbable. Cf. fn. 69 above. It is interesting, to say the least, that whereas the neo-Platonist Sallustius (fl. 363) who lived a century after Plotinus wrote a catechism περὶ θεῶν καὶ κόσμου (*Concerning God and the World*) to protect the youth of his day from the scourge of Christianity, Tillich would seem to avail himself of neo-Platonic categories in an attempt to apologize for the Christian faith. The catechism is sometimes but falsely attributed to another Sallustius (fl. 450-500), a cynic philosopher who is known to have heard the lectures of Proclus. Cf. *Dictionary of Greek and Roman Biography and Mythology*, 3 vols. (Boston: Little, 1849), vol. III, pp. 695 f.
104 Plato, *Timaeus*, 46; Aristotle, *Metaphysics*, 1072a; cf. *ibid.*, 1050a.
105 Proclus, *Elements*, 75-86, 97-100.
106 *Ibid.*, 87-96, 100-110.
107 *Ibid.*, 113-129.
108 *Ibid.*, 130-139.
109 *Ibid.*, 140-145.
110 *Ibid.*, 146.
111 *Ibid.*, 150-158.
112 *Ibid.*, 159-163.
113 *Ibid.*, 164-183.
114 *Ibid.*, 184-204.
115 *Ibid.*, 205-206.
116 *Ibid.*, 208-211.
117 Clement of Alexandria, *The Miscellanies*, The Anti-Nicene Fathers, Vol. II (Grand Rapids, Eerdmans, 1967),VI,iv, pp. 323 f.
118 Lactantius, *The Divine Institutes*, tr. M. McDonald, The Fathers of the Church, Vol. 49 (Washington: Catholic University Press, 1964), Book 1, Ch. 6, p. 32, and *The Wrath of God* in *The Minor Works*, tr. M. McDonald (Washington: Catholic University Press, 1965), Ch. XI, p. 87.
119 Cyril of Alexandria, *Adversus Julianum*, Patrologia Graeca, Vol. 76 (Paris: Migne, 1863), col. 548 B-C.
120 Augustine, *Concerning the City of God*, tr. H. Bettenson (Harmondsworth: Penguin, 1972), Bk. XVIII, Ch. 39, p. 814.

121 The *Corpus Hermeticum* now includes the fourteen books translated by Ficino, the *Asclepius*, and several other Hermetic manuscripts subsequently discovered. Cf. Walter Scott, *Hermetica*, 4 vols. (Oxford: Clarendon Press, 1924-1936), Vol. I, *Corpus Hermeticum*, pp. 17-48; also "Introduction", pp. 1-16. Cf. Frances Yates, *Giordano Bruno and the Hermetic Tradition* (London: Routledge and Kegan Paul, 1964), A.-J. Festugière, *L'Hermétisme* (Lund: Gleerup, 1948) and *La Révélation d'Hermès Trismégiste*, 4 vols. (Paris: Librairie Lecoffre, 1944-1954).
122 Plato and Aristotle had four: fire, water, earth, and air.
123 In Plato these causes are rest and motion. *Timaeus*, 57; cf. *ibid.*, 56-62 for the whole discussion. In Aristotle, it is a matter of the antithesis of opposites such as "density and rarity", "more and less", "excess and defect", or "potentiality and actuality". Aristotle, *Physics*, tr. P. H. Wicksteed and F. M. Cornford. (LCL,1934), I. v-vii,189a-189b; IV. vi, 213a.
124 Plato, *Timaeus*, 36 f. *Corpus Hermeticum*, tr. A. D. Nock and A.-J. Festugière, 4 vols. (Paris: Société d'Edition "Les Belles Lettres", 1945-1954), Vol. II, *Asclepius*, 10-11, pp. 308-310.
125 Ficino had the first fourteen of the treatises now included in the *Corpus Hermeticum*. He called the whole corpus, *Pimander*. *Corpus Hermeticum*, Vol. I, *Pimander*, I. 6-11, p. 8, line 14 to p. 10, line 9.
126 *Pimander*, I. 9, p. 9, lines 16-20.
127 *Ibid.*, I. 11, p. 10, lines 5-11.
128 *Ibid.*, I. 15, p. 11, lines 15-17.
129 *Ibid.*, I. 25-26, p. 15, line 15 to p. 16, line 15.
130 *Corpus Hermeticum*, Vol. I, *Hermes to His Son Tat*, V. 2, p. 60, lines 12-15.
131 *Ibid.*, V. 1-2, p. 60, lines 3-15.
132 *Ibid.*, V. 3, p. 61, lines 8-19.
133 *Ibid.*, V. 10, p. 64, lines 3-9.
134 *Corpus Hermeticum*, Vol. I, *Mind to Hermes*, XI. 10, p. 151, lines 14-19.
135 *Ibid.*, XI. 18, p. 154, lines 14-20.
136 *Ibid.*, XI. 20, p. 155, lines 9-12.
137 *Ibid.*, XI. 20, p. 155, lines 14-15.
138 *Ibid.*, XI. 7, p. 150, lines 9-17.
139 *Corpus Hermeticum*, Vol. II, *Asclepius to King Ammon*, XVI.2, p. 232, lines 18-19 to p. 233, lines 1-3.
140 *Ibid.*, XVI, 4-5, p. 233, line 23 to p. 234, line 3.
141 *Ibid.*, XVI. 7, p. 34, lines 11-16.
142 *Ibid.*, XVI. 17, p. 237, lines 11-14.
143 *Ibid.*, XVI. 17, p. 237, lines 14-16.
144 *Ibid.*, XVI. 17, p. 237, lines 16-19.
145 *Ibid.*, XVI. 18, p. 237, lines 20-24 to p. 238, line 1.
146 *Ibid.*, XVI. 19, p. 238, lines 1-6.
147 *Ibid.*, *Hermes to His Son Tat*, V. 11, p. 64. lines 16-19 to p. 65, lines 1-5.
148 For an introduction to the Hermetic writings, cf. Walter Scott, *Hermetica*, and Frances Yates, *Giordano Bruno and the Hermetic Tradi-*

tion.
149 Cf. Proclus, *Éléments de Théologie*, tr. Jean Trouillard (Paris: Aubier, 1965), p. 11.
Latin translations of some of the writings had been made at an earlier date. We have already mentioned that the Hermetic *Asclepius* was known among the early Latin Church Fathers as early as the late second or, in the case of Tertullian, early third century. Augustine, who read the text in Latin, knew of the writings in the late fourth or early fifth centuries. As for Plato, his *Timaeus* had been translated as early as the middle of the fourth century by Chalcidius (fl. 1st. half. 4th cent.). In the second half of the twelfth century, Henricus Aristippus (fl. 1156) had translated the *Meno* and the *Phaedo*. Proclus, too, must have been known to some extent in the West. In the sixth-century, for instance, Philoponos refuted his cosmological ideas. In the sixth century, the anonymous author "Pseudo-Dionysius", transposed Proclus' mysterious and celestial deity-filled hierarchical world into "Christian" terms. The Greek text of Proclus was translated into Latin in the ninth century. The translation was followed in the thirteenth century by a commentary written by William of Moerbeke (c.1215 - 1286), a contemporary of Thomas Aquinas. Thomas himself has but one citation, the first part of Theorem 100, that seems to have come directly from Proclus. His *Summa Theologica*, however, is replete with citations from Pseudo-Dionysius who "baptized" Proclus' neo-Platonic system into the Christian faith. It would also seem evident that Nicholas of Cusa was familiar with a Latin translation of at least that part of Plato's *Parmenides* in which Plato set out his own type of *Docta Ignorantia* along with the themes of the *one* and the *many*, *unity* and *difference* and "*the middle as equidistant from the extremes*". Thus, the mystery-laden, soul-filled world of Plato, made more mysterious and more soul-filled still by his neo-Platonic successors (perhaps as a counter to Christianity), formed a continuing theme in western thought from the sub-apostolic period right through the Renaissance and even beyond. Plato, *Parmenides*, 145, cf., *ibid.*, 134-135, 145-147, Cf. Cusa, *De docta*, I, Cap. 6, and Cap. XXVI. 86. Italics mine.
150 Marsilio Ficino, *Opera Omnia*, 2 vols. in 4. (Torino: Bottega d'Erasmo, 1959 [photo reproduction of Basel edition, 1576]), p. 806. English translations are taken from Paul Oskar Kristeller, *The Philosophy of Marsilio Ficino* (New York: Columbia University Press, 1943). Cf. *ibid.*, pp. 27 f.
151 Ficino, *Opera*, p. 1826. Kristeller, *Ficino*, p. 25.
152 Ficino, *Opera*, p. 1826. Kristeller, *Ficino*, p. 26.
153 Plato, *Timaeus*, 28-56.
154 Plato, *Laws*, X. 903.
155 Plato, *Timaeus*, 40-47.
156 A. N. Whitehead, *Adventures of Ideas*, (Cambridge: University Press, 1933), pp. 212 ff. Little wonder that Whitehead regards the Protestant

Reformation as an even more complete failure, in no way improving Catholic theology. *Ibid.*, p. 212. Whitehead's candid admission that he is not competent in the literature of the church may excuse his conclusion but hardly his temerity. *Ibid.*, pp. 211 ff.

[157] Ficino, *Opera*, p. 78. Kristeller, *Ficino*, p. 349.
[158] Ficino, *Opera*, pp. 93, 259, 275, 281, 333, 441, 1254 ff. Kristeller, *Ficino*, p. 164.
[159] Ficino, *Opera*, p. 103. Kristeller, *Ficino*, p. 165.
[160] Ficino, *Opera*, p. 148. Kristeller, *Ficino*, p. 165.
[161] Ficino, Opera, p. 282. Kristeller, *Ficino*, p. 134.
[162] Ficino, *Opera*, p. 282. Kristeller, *Ficino*, p. 167
[163] Ficino, *Opera*, p. 281. Kristeller, *Ficino*, p. 166.
[164] Ficino, *Opera*, p. 333. Kristeller, *Ficino*, p. 165.
[165] Ficino, *Opera*, p. 100. Kristeller, *Ficino*, p. 165.
[166] Ficino, *Opera*, p. 991. Kristeller, *Ficino*, p. 146.
[167] Ficino, *Opera*, p. 90 ff. Kristeller, *Ficino*, p. 61.
[168] Ficino, *Opera*, p. 218. Kristeller, *Ficino*, p. 62.
[169] "*Perficere illius est cuius est et facere*", Ficino, *Opera*, p. 1240. Kristeller, *Ficino*, p. 63, fn. 17.
[170] Ficino, *Opera*, p. 270. Kristeller, *Ficino*, p. 63.
[171] Ficino, *Opera*, p. 689. Kristeller, *Ficino*, p. 67.
[172] Ficino, *Opera*, p. 92. Kristeller, *Ficino*, p. 64.
[173] Ficino, *Opera*, pp. 250, 332. Kristeller, *Ficino*, p. 67.
[174] Ficino, *Opera*, p. 307. Kristeller, *Ficino*, p. 65.
[175] Ficino, *Opera*, pp. 174, 298, cf., *ibid.*, p. 231, Kristeller, *Ficino*, p. 69.
[176] Ficino, *Opera*, pp. 174, 298, cf., *ibid.*, p. 231, Kristeller, *Ficino*, p. 69.
[177] Aristotle, *De Anima*, (LCL, 1957), tr. W. S. Hett, III.ix, 432b.
[178] Ficino, *Opera*, p. 91. Kristeller, *Ficino*, p. 64. Here I would disagree with Kristeller that Ficino meant the statement to have a "universal and ontological sense" applied to the universe and of all its parts. Cf. *ibid.*, p. 69.
[179] Ficino, *Opera*, p. 1209. Kristeller, *Ficino*, p. 180.
[180] Ficino, *Opera*, p. 1208. Kristeller, *Ficino*, p. 176.
[181] Ficino, *Opera*, p. 248. Kristeller, *Ficino*, p. 130.
[182] Ficino, *Opera*, p. 981. Kristeller, *Ficino*, p. 116.
[183] Ficino, *Opera*, p. 238. Kristeller, *Ficino*, p. 116
[184] Ficino, *Opera*, p. 308. Kristeller, *Ficino*, p. 177.
[185] Ficino, *Opera*, p. 417. Kristeller, *Ficino*, p. 181.
[186] Ficino, *Opera*, p. 306. Kristeller, *Ficino*, p. 183.
[187] Ficino, *Opera*, p. 1251. Kristeller, *Ficino*, p. 187.
[188] Ficino, *Opera*, p. 307. Kristeller, *Ficino*, p. 190.
[189] Ficino, *Opera*, p. 193. Kristeller, *Ficino*, p. 193. Cf. Plato, *Gorgias, Dialogues of Plato*, Vol. 3, 525.
[190] Ficino, *Opera*, p. 375. Kristeller, *Ficino*, p. 197.
[191] Ficino, *Opera*, p. 675. Kristeller, *Ficino*, p. 197.
[192] Ficino, *Opera*, p. 417. Kristeller, *Ficino*, p. 196
[193] Ficino, *Opera*, p. 351. Kristeller, *Ficino*, p. 195.

146   The Renaissance, the Reformation and the Rise of Science

194   Ficino, *Opera*, p. 416. Kristeller, *Ficino*, p. 196.
195   Ficino, *Opera*, p. 417. Kristeller, *Ficino*, p. 196.
196   *Ibid.*
197   Kristeller, *Ficino*, pp. 27 f.
198   Ficino, *Opera*, pp. 871 f., Kristeller, *Ficino*, pp. 28 f
199   Ficino, *Opera*, pp. 991, 1673, 82. Kristeller, *Ficino*, pp. 146 f.
200   Ficino, *Opera*, p. 966. Kristeller, *Ficino*, p. 98.
201   *Ibid.* Emphasis mine.
202   Plutarch, *Moralia*, 15 vols. in 16 (LCL, 1957) *Concerning the Face Which Appears in the Orb of the Moon*, Vol. XII, 923 A. Cf. Nebelsick, *Circles of God*, p. 224, fn. 82.
203   Nebelsick, *Circles of God*, p. 221.
204   Ernst Zinner, *Entstehung und Ausbreitung der Coppernicanischen Lehre* (Vaduz: Topos Verlag, 1978), p. 160.
205   Marsilio Ficino, *Liber de Sole, Opera Omnia*, Ch. VI, VII, VIII, IX.
206   Nicolaus Copernicus, *On the Revolution of the Heavenly Spheres*, Great Books of the Western World, Vol. 16 (Chicago: Encyclopaedia Britannica, 1952), I. 10, pp. 526-528.
207   Cf. Nebelsick, *Circles of God*, esp. pp. 200-261 where the argument is stated in some detail.
208   Festugière, *L'Hermetisme*, pp. 15 ff.
209   Plato, *Timaeus*, 36; e.g. 1, $^4/3$, $^3/2$, 2, $^8/3$, 3, 4, $^{16}/3$, 6, 8 and 1, $^3/2$, 2, 3, $^9/12$, 6, 9, $^{27}/2$, 18, 27.
210   *Ibid.*, 35-36.
211   *Ibid.*, 55-56.
212   *Ibid.*, 36.
213   *Ibid.*, 36-37. Likewise, Aristotle reports that, for the Pythagoreans the elements themselves—earth, water, air and fire—were the manifestations of geographical figures. Earth manifested the cube; water, an icosahedron; air, an octahedron; and fire, the tetrahedron. Aristotle, *Meteorologica* (LCL, 1952), I.viii, 345b.
214   Plato, *Republic*, 525.
215   *Ibid.*, 526.
216   *Ibid.*, 527.
217   *Ibid.*,
218   *Ibid.*, 529
219   *Ibid.*,
220   *Ibid.*, 530.
221   *Ibid.*, 529.
222   *Ibid.*, 533.
223   Cf. F. E. Peters, *Aristotle and the Arabs* (New York: New York University Press, 1968), pp. 113-115.
224   Seyyed Hossein Nasr, *An Introduction to Islamic Cosmological Doctrines* (Cambridge: Harvard University Press, 1964), pp. 50 f. where he quotes from the *Rasa'il.*
225   Cf. *ibid.*, pp. 71 ff. Cf. Aristotle, *Physics* I.vi, 189a-189b.
226   Cf. F. Sherwood Taylor, *The Alchemists* (New York: Collier, 1962), pp.

82 ff.
227 Plato, *Philebus*, 29.
228 Cf. Festugière, *Révélation d'Hermès*, I, p. 215 where he refers to Marcelin Berthelot's *Origin de la Chemie* (1885); *Collection des Ancient Alchemists grecs*, 3 vols. 1887); and *Le Chimie au Moyen Age* (1883). The "art" itself, as indicated by the Hermetic documents, goes back to the early centuries of our era if not before. As Festugière points out, citing from the the works of Marcelin Berthelot (1827-1907) and others, there is a Greco-Roman period of alchemy as well as a history of the art among the Syrians, the Arabs, and the Latins of the Middle Ages. Since the difference between an atom of mercury and and an atom of gold is but a single electron, changing mercury into gold may now be within the realm of possibility for modern physics!
229 Cf. Festugière, *Révélation d'Hermès*, I, p. 218.
230 *Ibid.*, pp. 219 ff.
231 *Ibid.*,
232 Festugière, *L'Hermétisme*, pp. 16 f.

Chapter 3

THE REFORMATION AND THE RISE OF SCIENCE

Before there could be experimental science, the world had to become *demythologized* or *disenchanted* of its "immanent divinity". The spell of Aristotle had to be broken. As indicated in "The Christian Critique of Aristotle" above, Philoponos, Grosseteste, Roger Bacon, Duns Scotus, Ockham and Buridan represent those who had begun to show the inherent faults in the Aristotle's monolithic cosmogony. The disenchantment occurred when Aristotelianism was confronted variously with a combination of theological and scientific doctrines. Among the most important were (1) the doctrine of *creatio ex nihilo,* which entailed the unity of all creation, the heavenly as well as the earthly; (2) the identification of primary reality with individual entities; (3) the necessity of differentiating the different levels of truth and the use of inductive logic in reaching generalizations from the observation of particulars; (4) the law of parsimony; and (5) the theory of impetus which identified motion with that which moved. Under the pressure of these five considerations, Aristotle's enchanted house was so compromised that with the added weight of Renaissance thought, on the one hand, and of the Reformation on the other, Aristotelianism crumbled and a foundation was prepared upon which experimental science could build.

The introduction of Aristotle and the writings of ancient Greece to the West not only revived interest in the classical world, thereby initiating the Renaissance, but resuscitated interest in the sources of Christian thought as well. The rediscovery of the central significance of the biblical documents led to the Protestant Reformation and to a new orientation in the Christian understanding of the created universe. A new appreciation of the primacy of the grace of

God in the doctrines of creation, providence, and salvation brought with it a deeper understanding of the freedom of God in his relations with the creation, on the one hand, and of the freedom of the world in its creaturely differentiation from the Creator, on the other. The effect of this transformation was a two-fold gift of freedom: first, nature was emancipated, freed to be nature in its own created right; second, humankind was emancipated, freed to serve God and his intentions for creation. Humankind was freed to serve the world through exploration, investigation and technology. Basic to all of this was the doctrine of *creatio ex nihilo*. Creation was thus understood as God's sovereign act by which he gave what he had created the kind of integrity, stability and a regularity that allowed it to be accessible to rational knowledge and control.

BIBLICAL FAITH AND THE RISE OF SCIENCE

Creatio ex nihilo
For Israel history and creation are of a piece. Yahweh, the God of Israel, whose *name* is "majestic in all the earth", and whose "glory is chanted above the heavens", is sovereign over both (Ps. 8:1). He is *El Shaddai*, God Almighty, the Lord of history who appeared to Moses as he had to Abraham, Isaac and Jacob (Ex. 6:2-3). According to the creation accounts in the Book of Genesis, God created the world as his first great act and with that act history began. Though Old Testament scholars generally agree that the doctrine of *creatio ex nihilo* is not explicit in the Genesis account, nevertheless, the point of the story is that God brought order out of chaos.

As the Hebrew term, *bara* (to create) — which is reserved exclusively for God's creative activity — indicates, God brings the world into being by his Word and is sovereign over and independent of his creation. In coping with the pressures of Hellenistic culture, pre-Christian Judaism interpreted

God's sovereign act of creation as creation out of nothing. When in line with this tradition the Apostle Paul could say of God that he "calls into existence the things that do not exist" (Rom. 4:17), he set the pattern for a Christian understanding of *creatio ex nihilo*. Patristic theology then gave this understanding explicit doctrinal formulation and the result was that the concept of God's sovereignty over and independent of creation could be transferred into Hellenistic thought constructs.[1]

It was in this sense that Reformation theology placed the biblical account of creation over against the Graeco-Roman *deus sive natura* world-view. In doing so it gave the *coup de grâce* to the Aristotelian synthesis of God and the world. Now God the Creator was understood as the sovereign, active Lord of history and creation. Having been brought into being by the Creator in utter differentiation from himself as an orderly independent reality, creation exists in contingent freedom.[2]

It is important at this point to realize that the biblical writers were not inventive cosmologists. They took the three-storied structure of the universe for granted (Ex. 20:4). They pictured the earth as a flat surface, ruffled with mountains and run through by rivers. They saw the firmament, spread out by God "like a curtain" and "like a tent to dwell in" (Is. 40:22). It was a dome-like structure that held back the heavenly waters (Gen. 1:8, Ps. 148). They believed that the flat earth was founded on pillars which were sunk in the waters below the earth (Ps. 24:2, 104:5).

The abiding innovation afforded by the biblical faith, however, was the concept of Yahweh who in majestic but active sovereignty is exalted above the whole of creation and controls it (Is. 40:22). Hence, even though the myth of the god Marduk (the Babylonian god of order) over the monster Tiamat (the goddess of the sea and hence of chaos) is reflected in Yahweh's struggle with the sea monsters Rahab or Leviathan (Job 9:13, Ps. 74:13, Is. 27:1, et al.), Yahweh remains Lord of the totality of creation. He not only set the boundaries of the sea,[3] he holds in check the waters which surround the earth and those that are below it lest they burst forth and engulf the land (Gen. 1:6, 7:11). As Jaki reminds us, the Babylonians "remain trapped in the disabling sterility of a world-view in which not reason ruled but hostile willfulness". They believed "they were a part of a huge, animistic cosmic struggle between chaos and order, the final outcome appeared to them unpredictable and basically dubious".[4]

In contrast the biblical creation accounts emphasize that both the foundations of the world and the forces of nature were controlled by Yahweh. He thus employed the forces of nature for his own purposes. He caused "natural" plagues to overcome the Egyptians. He used winds and the sea to clear a path through the Red (Reed) Sea, so that the children of Israel could escape Pharaoh's army (Ex. 15:21). He performed signs and wonders in the wilderness and at Sinai. At the battle of Megiddo Yahweh caused the stars to fight on the side of Israel and with the torrent of the River Kishon he swept the enemy away (Jg. 5:20-21).

Little wonder that the Jews, whose first three of the Ten Commandments (Ex. 20) were explicit injunctions against recognizing the gods or goddesses that were identified with humankind or nature, rejected the mythic cult of their neighbors including that of their Babylonian conquerors even as they celebrated the sovereignty of Yahweh over all creation (Ps. 24, 27, 93-100). Yahweh, the sovereign Lord of history, had shown his power over both political powers and those of nature is rescuing Israel from Egypt. Therefore:

> You shall have no other gods before me. You shall not make for yourself a graven image, or any likeness of anything that is in heaven above, or that is in the earth beneath, or that is in the water under the earth; you shall not bow down to them or serve them; for I the Lord your God am a jealous God. (Ex. 20:3-5)

According to the creation account in Genesis 1:1-2:4a, *"in the beginning"*, i.e., independent of any previous power, Yahweh's spirit moved over the waters. By his Word, by his active self, he called light into being and separated it from darkness. In the same way he created the firmament to separate the waters above from those below. He made land, separated the waters on earth into seas and called forth plant life as an ordered creation. Next he called the stars, the sun, and the moon into being, put them where they belonged, and assigned them their separate functions. Then came the fish, the sea life, and the birds. Nearing the end of creation, Yahweh ordered domesticated animals, reptiles, insects and small quadrupeds "the creeping things" — and the wild beasts to come forth all in their proper places. Lastly, after taking counsel, he created humankind in his *image* to be stewards of creation. In creating humankind in his image, Yahweh delegated his authority over creation to humanity, with the command to rule it and by caring for it, to keep it in order (Gen. 1:1-28). Thus, to quote Reijer Hooykaas:

> He is truly all-powerful; He is not opposed by any matter that had to be forced into order, and He did not have to reckon with eternal Forms; His sovereign will alone created and sustains the world.[5]

In this way the biblical account assures us that Yahweh is not subject to other powers neither does he contest with them; rather all powers in heaven and on earth are dependent upon Yahweh the Creator (Ps. 8:6-7, Is. 44:24). Creation in turn testifies to Yahweh's wisdom, steadfast care, incomparable majesty and glory (2 Kg. 19:15, Neh. 9:6, Ps. 104:24, 136:4-9, Is. 40:25-26, Jer. 10:6-7). "The heavens declare the glory of God and the firmament proclaims his handiwork" (Ps. 19:1). Therefore:

> Let all the earth fear the Lord,
> let all the inhabitants of the world stand in awe of him!
> For he spoke, and it came to be;
> he commanded, and it stood forth. (Ps. 33:8-9)

In the New Testament we discover the same radical message of the sovereignty of God spelled out in Christological terms. This took place in the midst of the spirit-filled enchanted world of the Hellenistic cults and the Hellenistic culture. According to the New Testament Christ represents the "Almighty", the *pantocrator*, the "all-ruler" (2 Cor. 6:18; cf. Rev. 1:8, 4:8, 15:3, et al.). As Yahweh is sovereign over chaos, holds the waters back, and creates the heavenly bodies to serve him, so Christ has overcome the powers and principalities (Rom. 8:38). Hence, "at the name of Jesus every knee should bow, in heaven and on earth and under the earth" (Phil. 2:10). In him through whom are all things (1 Cor. 8:6), all things cohere and hold together (Col. 1:17). Eventually all things are to be united in him (Eph. 1:9-10) and even now he upholds the universe by his power (Heb. 1:3).

According to Plato, the demiurge who is the "father" or "begetter" of the universe will not stoop to make the world or humankind. Rather, he delegates this lesser task to the "world soul" or "immanent world reason".[6] In Aristotle the generation of nature and its creatures comes about by the reproduction of eternal forms.[7] By contrast, the Prologue to the Gospel of John uses Greek terminology to express Hebrew thought forms by directly reflecting the concepts of Genesis wherein Yahweh creates and orders creation apart from himself by his Word. Yet he creates it in a contingent relationship to himself. In transferring the meaning of the Hebrew term, *dabar*, "word", through which the world is created, into the Greek *logos*, "word", through which *all things* came into being (Jn. 1:3), the author of the Gospel of John offers the Hebrew gospel to the Greeks.

> In the beginning was the Word, and the Word was with God, and the Word was God. He was in the beginning with God; all things were made through him, and without him was not anything made that was made. In him was life, and the life was the light of men . . . The true light that enlightens every man was coming into the world. He was in the world, and the world was made through him, yet the world knew him not. (Jn. 1:1-4, 9-10)

The creating word, although separate from creation, is also "in the world". The "logos", as Arthur Peacocke points out, is "a profoundly fruitful conflation" of the Old Testament "word of God" or *dabar* and the Hellenistic-Jewish idea of God's wisdom or *sophia* "as an immanent power of God within the world and in man".[8] It is this "creative power," as C. H. Dodd puts it, "by which the universe came into being and is sustained".[9] One should be careful, however, lest in adopting Philo's (c. 20 B.C. - c. A.D. 50) idea that the *logos* is "the thought" of the "ultimate Deity," one adopts the Platonic idea that the *logos* belongs to creation as "the principle reality of the universe". Dodd seems to do this.[10] In biblical terms, God himself as *logos* is not a world soul or eternal form. Nevertheless he is responsible for creation both in terms of its beginning and for its continued existence.

The identity between Yahweh and Christ, presumed in θεός - λόγος (God-Word) identity of John 1 is continued by the author of the Book of Revelation. The author harks back to Second Isaiah where Yahweh knows the beginning and the end. Hence he is the God of the whole of history and the whole of creation. In the same way, the Book of Revelation harks back to "the Lord of hosts" of the prophet Isaiah and identifies Christ with him by calling Christ "the Alpha and the Omega, the first and the last, the beginning and the end" (Is. 44:6, Rev. 1:8, 4:8, 15:3, 22:13).

In this way the identification of the God of Scripture as sovereign denigrates the gods of the Greeks to mere creations of humanity who, like humankind, are of creation. At the same time nature is relieved of the divine qualities that the Greeks attributed to it. The disenchantment of the world, or in Hooykaas' words, "the de-deification of nature,"[11] is thus accomplished by the Word of Yahweh, the God of Israel, the Creator of heaven, earth, the sea and all that is in them. In Christian terms, this process continues with Jesus Christ, the *logos* made flesh, the *Alpha and the Omega*, who is understood as the ruler of all, the one who has overcome all "powers and principalities". The world, understood as a reality contingent upon God alone, was thus allowed a unity, a stability and an order.[12]

With the conception of *creatio ex nihilo*, competitive and intervening powers, demiurges and mini-deities of all kinds were banished. In this way all creation as a whole, that of the heavens and of earth as well as the forces governing them, was secularized. The concept of the world as an independent cosmic order apart from God, but in a contingent relationship to God, invalidated the intervening, capricious Greek divinities.[13]

In this context the Protestant Reformation, with its emphasis upon the Word of God, may be seen as a re-emphasis on the monotheism of God, the Creator who is represented by Jesus Christ, the *Word made flesh*. Again, God is God and the world as the world is secular. As Torrance points out, it is with the Reformation that there is a rejuvenated focus on the "Godness of God" and the "naturalness of nature". This has implications for both theology and science. It is the sheer differentiation between God and nature which definitively overcomes both the sacramentalism of Augustinianism and the *aeternitas mundi* (the eternality of the universe) of Aristotelian philosophy.[14] Thus the concept of *deus sive natura* is finally dispensed with.[15]

Hence, contrary to Whitehead who emphasizes the role played by medieval theology in the development of science,[16] it must be pointed out that "the rationality of God, conceived as the personal energy of Jehovah"[17] became operative in theology with the Reformation. It is evidently true, as Whitehead insists, that the faith in the harmony and reliability of creation, unique to the thought of Europe, is a direct result of theological thought. This faith, however, was quite antithetical to the understanding of the interpenetration of God in the world which was basic to the very structure of medieval theology and its doctrine of *analogia entis* (the analogy of being by which the nature of God could supposedly be ascertained from the nature of creation). Thus although the medieval theology of Thomas Aquinas, on which Whitehead bases his argument, denied Aristotle's insistence on the eternality of the world, with Thomas, the God of the Bible continued to be in tension with Aristotle's God who was first cause and who interpenetrated nature in the form of secondary causes.[18] As Torrance has pointed out:

> The infinite difference between the being of the Creator and the being of all created reality implies not only the rejection of the notion that God, even as First Cause, necessarily and logically belongs to what he has made, but also the all-important concept of contingent rationality inherent in created reality.[19]

Hence, a proper sense of the biblical transcendence of God over nature and the concept of nature as free from God but contingent upon him is not as distinct in Thomist theology as would have been the case had Thomas allowed his biblical theology to overcome his Aristotelian predilections. Wilhelm Dilthey (1833-1911) underscored this point when he argued that it was the dynamic created by Thomas' marriage of Aristotle's unmoved divine first-cause to the biblical active Creator and Redeemer that eventually tore the fabric of medieval metaphysics apart.[20]

The conception of the intelligibility of creation as creation, then, stands in contrast to, as much as in continuity with, medieval theology. In overcoming the negative aspects of Thomism, the mental outlook that both culminates in and issues from the Reformation effectively liberated creation to be itself. Consequently scientific thinking as we understand it was allowed to come to the fore. Freed from implacable divine causes, creation could begin to be known in terms of a rationality appropriate to it. Thus Whitehead's argument that science depends on the "inexpungeable belief that every detailed occurrence can be correlated with its antecedents in a perfectly definite manner, exemplifying general principles", reflects to a greater degree postmedieval rather than a medieval conceptuality.[21]

For science to begin, nature had to be seen as dependable, intrinsically worthwhile and knowable. It had to be understood in terms of a contingent rationality appropriate to it rather than in terms of a divine rationality that penetrated it and, as in Whitehead, became *concretized* in the manifestations of creation. Thus, we could not agree more with Whitehead's thesis that the dependability upon which science relies requires the belief in the rationality of the universe. Rationality depends upon faith in the rationality of God.[22] Nevertheless, contrary to Whitehead, God's rationality is not to be identified with any part of that of the world.

It is in this context, then, as Jaki has pointed out, that the words of the French astronomer Pierre Laplace (1749-1827) to the effect that science has no need for a "God-hypothesis",[23] reflects "a thoroughly sound scientific attitude".[24] Equally profound and equally difficult for pious ears to hear is Dietrich Bonhoeffer's (1906-1945) statement that Pierre Laplace's injunction against "the working-hypothesis of God" also reflects a thoroughly sound theological attitude.[25] God is transcendent to creation and therefore not an hypothesis that one can read off the face of creation in any way at all.

The universe is thus free and contingent upon the rational freedom guaranteed by God. It exists in "contingent freedom" and is characterized by its own order.[26] The Old Testament witnesses again and again to the faithfulness of God and the consequent dependability of creation. God's decree has established "the ordinances of the heavens" (Job 38:33) and the bounds of the deep (Ps. 104:9). He gives rain in its season (Jer. 5:24) and maintains the fixed order of sun, moon, stars, and sea (Jer. 31:35-36). The faithfulness which God showed at creation, he keeps forever (Ps. 146:6). While the earth remains, seedtime and harvest, cold and heat, summer and winter, day and night, shall not cease (Gen. 8:22).

Jaki makes the same point from a Christological point of view. "The Christian certitude about the order of nature, about man's ability to investigate its laws, owes its vigour to the concreteness by which Christ radiated the features of God creating through the fullness of rationality which is love."[27] Thus the Apostle Paul who knows Christ to be "the image of the invisible God" (Col. 1:15) can express his confidence in the basic subordination of the whole of creation to God in Christ, a subordination which implies a contingency of creation upon God.

> For I am sure that neither death, nor life, nor angels, nor principalities, nor things present, nor things to come, nor powers, nor height, nor depth, nor anything else in all creation, will be able to separate us from the love of God in Christ Jesus our Lord.
> (Rom. 8:38-39)

It is from this perspective, a perspective that we have yet fully to understand, that we can appreciate Torrance's incisive insight that "the orderliness of what is contingent seems to have arisen out of Christian theology".[28]

We know that the pre-Socratic Ionians, such as Thales,

Anaximander, Anaximenes and Anaxagoras were on their way to recognizing the value of nature as nature. However, with the Pythagoreans, Plato, and Aristotle, Greek conceptions of God and nature had the effect of impregnating the world with both deities and divine first and final causes.[29] The result was that their conceptions of nature and of God were muddled. The confusion, carried to the West first by the neo-Platonic sacramentalism of Augustine and then by the writings of Aristotle, was so damaging that neither God nor nature could become "subject" to scientific inquiry. With the help of the biblical theology of the Reformation, the mind was enabled to conceive of God as both Creator and sovereign Lord and to think of nature, as independent of God and yet as unceasingly contingent upon his creative power.[30] Rather than being *God-like*, or rather than being understood in terms of sacramental signs of God's presence, the world could be understood as having a reality of its own. Instead of being subject to the incessant but unpredictable divine intervention of some *deus ex machina*, the world could be regarded as having its own substantiality under God and be understood in terms of a rationality which itself is a part of created reality.

Thus the world is neither the manifestation of the divine under the exigencies of finitude as F. W. J. von Schelling (1775-1854) argued, nor is it the arena in which God concretizes himself in existence, as Whitehead would have it.[31] Neither is nature *within* God in a "panentheistic" fashion.[32] Rather, God is to be understood as transcendent "over" nature. At the same time, God providentially *maintains* the world in independence and freedom from himself.

As Hooykaas points out, because "the idea of a divine Creator implies the absolute dependence of created things upon him and also their total differentiation from him",[33] nature is given freedom to operate according to God-given and God-sustained rational structures appropriate to it. God is God and nature is nature. As such, nature can be treated as a knowable, accessible, reliable subject matter that may be examined, experimented on, understood, used, and molded. In short it is the realm of reality with which both science and technology can be effective. Belief in a rational Creator and belief in a universe subject to rationality and contingent upon the Creator go hand in hand. When the biblical theology of creation and Christ was assumed the God of biblical revelation replaced the pantheon of the Greek gods. Consequently the world was freed for scientific investigation.

While the Reformation certainly opened up people's minds for modern science and technology, it cannot be claimed that the Reformers themselves had a monopoly of influence in this direction. As Jaki points out, apart from the anticipations which we have already noted, in 1277 Étienne Tempier (d. 1279), Bishop of Paris, had already condemned a number of Aristotelian propositions on the ground that they limited the power of the Creator. Pierre Duhem (1861-1916), an historian of science, regarded the ban as being of such importance that he pointed to the year 1277 as the date that modern science began.[34]

Apart from the question of whether or not ecclesiastical condemnations against free inquiry ever promoted knowledge, subsequent history indicates that, well-intended though it was, the prohibition was but a futile attempt by ecclesiastical authorities to stem an inevitable tide. According to A. C. Crombie, in 1255 twenty-two years prior to this ban the University of Paris set Aristotle's most important metaphysical and natural works as subjects for examination.[35] The interdiction of 1277 seems to have been directed against the general encroachment of pagan Averroism on Christian theology. It condemned 219 propositions in all. As a number of these were from the writings of Thomas Aquinas, who more than any other was responsible for making Aristotle *de rigueur* for the West, the ban was, at best, temporarily effective. In 1323 Thomas was canonized, and his statements were removed from the condemned list. This indicates that by then Aristotle was no longer seen as a threat.

The tide of Aristotelianism continued to overwhelm European theology for the next three centuries. At the beginning of the seventeenth century Aristotelianism had become the measure of orthodoxy to such an extent that in 1616 Copernicus' *On the Revolution of the Heavenly Spheres* was condemned and put on the *Index of Forbidden Books* for contravening the Aristotelian teaching that the earth stood still. In 1633 Galileo's writings were put on the *Index* for the same reason. It is also of some consequence to note that when Descartes received news of Galileo's condemnation, rather than publish anti-Aristotelian views, he destroyed his own heliocentric treatise which he had entitled *Monde (Cosmos)*.

There is little doubt that from the end of the sixteenth century onward Aristotelianism was adhered to more fervently than during the thirteenth, fourteenth and fifteenth centuries. In 1624, more than a hundred years after the inception of the Reformation, the Parliament of Paris threatened with the death penalty anyone who maintained a doctrine contrary to Aristotle. Aristotle's thought, so basic to the medieval synthesis, remained entrenched in certain circles long after the bloom of medieval theology withered away.

In addition recent history has shown that the struggle to keep the secular secular and hence to renounce the concept of *deus sive natura* for the sake of science must be fought again and again. One needs only to allude to Whitehead's and Tillich's versions of an immanent relation between God and nature. The most disastrous example is the theology of the Deutschen-Christen of Nazi Germany. Howe reminds us that the development of this theology illustrates that even so-called "Christian cultures" are not immune to the confusion between God and creation, and that this can have fateful and destructive consequences.

> The old middle eastern Semitic culture (Babylonian) celebrated a curious resurrection during the rise of National Socialism in Germany. However, instead of the Babylonian ground being holy, it was the German ground. Holy was no longer the Babylonian man, but the nordic blood. And instead of the Babylonian king, it was the god-sent *Führer* who knew how to interpret that which he called "providence". He was the guarantee of divine power and presence.[36]

The chaos that resulted should have been predictable. It is hardly a coincidence that the Jews, who are even less tempted to mix God with the world than are Christians, were the primary victims of this confusion of the sacred and the secular. Both science and civilization depend on keeping the secular secular. This, in turn, depends on our continual recognition of Yahweh, the God of Jews and Christians, and our steadfast knowledge that creation is his creation *ex nihilo:* "Thou shalt have no other Gods before me."

"In the very center of God's revelation", says Howe, "stands the mystery-laden divine name, Yahweh" ("I will be who I will be" or "I am who I am"). He goes on to say that the differentiation between the biblical name of God and the Greek concept of being is immediately apparent. The concept of being in the name of God does not imply a metaphysical being in and for itself; rather, it refers to the reality of the Living God whose mighty acts are revealed in salvation history. The emphasis is placed not upon passive but upon active existence. God's promises include his whole creation. Both nature and human nature are caught up in his creating, sustaining and redeeming lordship, so that in the end both nature and humanity are destined for redemption and renewal. Hence, the author of the Book of Revelation envisages "a new heaven and a new earth" (Rev. 21:1); "he who sat on the throne said, 'Behold, I make all things new'" (Rev. 21:5).[37]

Covenant Relationship
God is God and creation is creation. Of equal importance to the development of science is the relationship of God to humankind and the relationship of humankind to creation. The first defines the position of humankind vis-à-vis creation. The second sets out their responsibility. Again, the creation accounts of Genesis 1:1 ff. (which, as stated above, rest on older accounts in the Psalms and other Old Testament writings) provide the relevant categories. As Jaki puts it, "The story of Yahweh making the world and preparing it for man, who is God's special handiwork, was obviously

told many times before receiving its final written form and was certainly repeated on countless occasions following its definitive phrasing".[38]

In contrast to the Babylonian and Greek conceptions of divinity and humanity, Genesis 1:26 names humankind as God's regent. As God's stewards humanity is given the task of ruling creation and *using* it for its own purpose. Being created in the image of God and being given responsibility for creation, humanity especially as Reformation theology has expressed it, exists as God's "covenant partners". Adam and Eve are not divine but *godlike* in being responsible for God's creation. As God's partners they have a two-fold relationship. On the one hand, they are related to God *to whom* they are responsible; on the other, they are related to nature *for which* they are responsible. Thus, the *covenant-partner relationship* is not only an expression of the position of humankind in reference to God, it is also an expression of the position of God in apposition to nature. Humanity mediates between God and nature. Humanity and nature together, as both Karl Barth (1886-1968) and T. F. Torrance have pointed out, must be seen within the realm of grace where God is understood as being concerned for the salvation of both humankind and nature.[39]

From a biblical perspective, the whole of history and the whole of nature are bound up with humanity. In that it is through humanity that God expresses his will for creation, even the act of creation itself is best interpreted in relation to the covenant and humankind. Yahweh is both Creator and Lord over his creation. In Israel's pre-history he had called Abram out of Ur, bound himself in covenant relationship with Abram and promised him that through him all the nations would be blessed (Gen.12:3). He rescued Israel from Egypt, proclaimed himself to be the people's liberator and regulated their lives by giving them the Ten Commandments (Ex. 20:1-17).

The creation account of the second chapter of Genesis is a rendition of events which moves quickly to its climax, the creation of Adam. Adam is made the curator of "the garden", *i.e.*, the rest of creation, so that creation might be cared for. "In the day the Lord God made the earth and the heavens . . . there was no man to till the ground" (Gen. 2:4-5). And in the shortest possible time after the earth had been watered so the dirt was moist for moulding: The Lord God formed man of dust from the ground, and breathed into his nostrils the breath of life; and man became a living being. (Gen. 2:7) As Jaki has said, this "includes the preparation of the whole of nature for him".[40]

The justification for Barth's understanding of the covenant as "the inner ground and form of creation", on the one hand, and creation as "the outer ground and form of the covenant", on the other,[41] is founded on this biblical basis. Barth's scheme, nonetheless, does not include the redemption of the whole of creation within the covenant relationship. This lays too great a stress on *anthropology*, the doctrine of and place of humanity in creation, to the detriment of *cosmology*, the doctrine of nature. The very fact, however, that from the creation accounts onward humankind and nature are seen as different but interrelated aspects of creation (the totality of which is under God's care and in need of salvation) indicates that *cosmology*, the doctrine of the cosmos, must be stressed along with *anthropology*, the doctrine of humanity. Both humankind and nature should be understood to be in a covenant relationship with God.

In both the Genesis creation accounts, humankind is responsible for nature. In the account in Genesis 1, as stressed above, Adam and Eve are mandated to subdue the earth and to dominate it (v. 28). They are to take charge of nature and nature's creatures and care for them, to keep the wild animals separate from the domesticated ones, and to prevent the latter from trampling the plants. The plants, in turn, are to be used for food for both humans and animals (vs. 29-30).

In Genesis 2 there is an intimate solidarity between humankind and nature. As mentioned above, humanity is created from the very stuff of which nature is "made". "The Lord God formed man out of the dust of the ground" (v. 7). God then set Adam in the midst of the garden "to till it and keep it" (v. 15). Adam is designated as the steward to care for nature. Because he cares for it, nature will care for him. He is to eat of the trees of the garden (v. 16). By naming the beasts and the birds he expresses his power over them and their dependence on him.

As the second creation story (Gen. 2:4 ff.) unfolds and sin enters the picture, the original and in a proper sense "natural" relationship between humankind and nature is disturbed. Sin disturbs not only the harmonious relationship between God and humankind but equally important, as far as science is concerned, between humankind and himself. Sin even causes the relationship between humankind and nature as well as the relationships between the individual components of nature to be disturbed and distorted. The result is disharmony, estrangement, pain, and suffering on all sides. The serpent, damned to go on its belly and eat dust, is designated the enemy of humankind. The woman must now endure pain during childbirth and to be subordinate to her husband. Even the ground is condemned to bringing forth thorns and thistles. And Adam, who must weed them out, is cursed with such toil that he shall sweat even while he eats until he "returns to the ground" from which he was taken (Gen 3:14 ff.). Because disharmony and disobedience reign, Cain's murder of his brother Abel again affects the fruitfulness of the soil. "When you till the ground, it shall no longer yield to you its strength" (Gen. 4:12). The earth, in fact, becomes so corrupt, so filled with violence, and humankind becomes so rebellious that God decides to blot out his creation and start anew (Gen. 6:7).

The flood comes; but not even the flood alters God's original intention. When the ark lands and the creatures are released, the original mandate is repeated:

> Be fruitful and multiply and fill the earth. The fear of you and the dread of you shall be upon every beast of the earth and upon every bird of the air, upon everything that creeps upon the ground and all the fish of the sea; into your hand they are delivered. (Gen. 9:1-2)

Again humankind are put in charge and again, nature is as dependent upon them as they are dependent on nature. This time, in contradistinction to Genesis 1 and 2, in Genesis

9 not only the plants but also the animals are designated as food.

> Every moving thing that lives shall be food for you; and as I gave you the green plants, I give you everything. (v.3)

As Yahweh is the lord of all creation there is also restriction. "Only you shall not eat flesh with its life, that is its blood" (v. 4). Animals, like humankind, have the "breath of life". They have a *nephesh* which translates as "soul". Consequently the blood, which was thought to be the seat of the soul, is not to be eaten. In fact, God demands that the animals be protected. He will require a reckoning of "every beast" as well as of "every man" (v. 5). And as Noah is reminded that "God made man in his own image" (v. 6), the emphasis, as in Genesis 1, is on responsibility for the earth and its creatures. Hence in the covenant that God makes with Noah and his descendants, he also includes "every living creature"—the birds, the cattle, and all of the beasts (vs. 10 ff.). Since all are important to God:

Never again shall all flesh be cut off by the waters of the flood and never again shall there be a flood to destroy the earth. (v. 11)

Now Noah, like Adam and Eve before him, becomes a gardener, a steward of creation, a "tiller of the soil" who plants a vineyard (v. 20). To till the soil, *abad*, is to "serve the earth". It is to cooperate with creation so that creation in turn can support humankind.

This symbiosis between humankind and nature, so important to both faith and science as well as technology is seen as moving toward its proper perfection in Isaiah's vision of harmony and peace. With the total re-establishment of God's rule, there will be the full reconciliation of the whole of creation.

> The wolf shall lie down with the lamb,
>   and the leopard shall lie down with the kid,
> and the calf and the lion and the fatling together,
>   and a little child shall lead them.
> The cow and the bear shall feed;
>   their young shall lie down together;
>   and the lion shall eat straw like the ox.

> The suckling child shall play over the hole of the asp,
>   and the weaned child shall put his hand on the adder's den.
> They shall not hurt or destroy in all my holy mountain;
>   for the earth shall be full of the knowledge of the Lord
>   as the waters cover the sea. (Is. 11:6-9)

The New Testament, too, sees the salvation of humankind and the salvation of creation as interrelated and interdependent. Christ is the new "Adam", the new man who represents life since Adam, and because of his sin, represents death. He is the first fruit of the new order (1 Cor. 15:23 ff.), the order which is being brought to birth in the power of the spirit (Jn. 3:5). As the resurrected Lord, Christ manifests the new creation in himself and is the very foundation of the new creation of all things (2 Cor. 6:18; cf. Rev. 1:8, 4:8, 15:3 et al.). "All things have been put under his feet" (Rev. 15:25, Eph. 1:22). In him the totality of reality finds its focus and its end.

> He is the image of the invisible God, the first-born of all creation; for in him all things were created in heaven and on earth, visible and invisible, whether thrones or dominions or principalities or authorities—all things were created through him and for him. (Col. 1:15-16)

This solidarity of humankind and nature in the covenant relationship is the foundation of the salvific strategy wherein the Apostle Paul sees the whole of creation anticipating the salvation to come.

> We know that the whole of creation has been groaning in travail together until now; and not only the creation, but we ourselves, who have the first fruits of the Spirit, groan inwardly as we wait for adoption as sons, the redemption of our bodies. (Rom. 8:22-23)

When humanity is redeemed, creation which has been subject to futility (Rom. 8:20) will "itself be set free from its bondage to decay and obtain the glorious liberty of the children of God" (Rom. 8:21). In the meantime, creation itself sees the pattern of its hope in those who manifest the initial characteristics of redemption.

The main impact of biblical anthropology in this regard is the pronouncement that humanity *under God* is free to have dominion over rather than be dominated by creation.

This benevolent dominion is for the sake of *both nature and for humanity*. Within God-given limits, under God and in relation to God, humankind are capable of recognizing themselves as responsible for both themselves and the world. They are enabled to know themselves and nature, to enter into self-reform and the reform of society, and to exercise control over that part of the world of nature under their dominion.

There is no doubt that humankind is continually tempted to be creation's sole master. Martin Heidegger's (1889-1976) statement, "He [humanity] is no longer satisfied to be the shepherd of being but wants to be the lord of that which is, no longer the gardener of nature but nature's lord and its possessor",[42] shows the distortion of an understanding which knows power but not responsibility. Howe explains this distortion in terms of "false objectification".

> In objectivizing the world man has so grasped it that he has made it into a machine. There are certainly no more nereids and nymphs. Now, however, as Hegel and Marx have put it, all that is, is now only manipulatable material and the whole world stands under the lordship of the logical reason in man, the logical eye.[43]

Hence, rather than agreeing with Lynn White who blames the West's current ecological crisis on "orthodox Christian arrogance toward nature" which follows from the imperative of the Genesis command,[44] we must understand the command as does Claus Westermann who insists that human dominion of nature is to be compared to that of a shepherd. It is caring rather than exploitive.[45] Norbert Lohfink's insistence on the oriental concept of the ruler as shepherd points in the same direction.[46]

Thus, the original intent of the Genesis account and the identity of humanity and nature integral to the covenant relationship has a three-fold implication for the understanding of the order of creation. First, it spells out the Lord-servant relationship between God and humanity. Second, it shows the dominance-dependence relationship between humanity and nature. Third, it indicates the original harmony within nature itself. The order is established to insure that the interdependence of humanity and nature continues to be mutually beneficial.

The fact that humankind's dominion of nature has been distorted leading to exploitation should not mislead us into denying the ordering of nature as it is proclaimed in scripture, nor should we shirk our responsibility within the ordering-caring scheme.[47] Rather, it would seem that we must understand with J. Robert Nelson that the unwarranted exploitation of nature is due to sin and selfishness.[48] This sinfulness affects nature itself. The fact that the inner harmony of nature is disturbed would seem to call into question the harmonious rationality of nature *qua* nature in its present state. Such thinkers as Einstein were convinced that this harmony existed. On this basis he continued to reject the non-harmonious aspects of nature as understood by quantum mechanics in general and by Werner Heisenberg's (1901-1976) Principle of Non-Determination and Niels Bohr's (1885-1962) Theory of Complementarity in particular.

From a biblical perspective, natural science, the knowledge of nature and its application, technology, may be understood as a new appreciation of the Genesis command to have dominion over nature not for the sake of exploitation but for the sake of the whole of God's creation. It was due to human folly, Francis Bacon claimed, that the human race lost its dominion over the created world in the Fall.[49] Bacon explained that even sinful humanity could assert command over nature if they worked sufficiently hard to secure it. Bacon, in fact, was convinced that it was a part of our *religious duty* to build God's kingdom through *sciencia*.[50]

Likewise, our contemporary, Klaus Koch argues that in the account of creation of Genesis 1, the whole program of modern natural science and technology lies, '*in nuce*', before us![51] If so, and considering the crisis of the world in which we find ourselves, we should not see science and technology as foreign to humankind. On the contrary we should understand science to be a proper implication of the faith and a legitimate and practical extrapolation of the Genesis command.[52] For only the proper exercise of science and technology will enable us to approximate the goal of keeping nature in order and re-ordering it toward its nature. This implies that we should understand nature in the light of a rationality of redemption. From the biblical perspective redemption is not intrinsic but is offered as a gift. Were it to be redeemed then creation might again be pronounced "good", even "very good" (Gen. 1:31).

According to our analysis, there is a coincidence between the faith, confidently centered on the Christian doctrines of creation and covenant and the rise of modern science. The confidence necessary for engaging in the scientific enterprise can in fact be seen as directly related to the teaching of the sovereignty of God expressed in both doctrines. As the understanding of the responsible role of humanity in creation depends on faith, so too, do our understandings of history and time as well as our appreciation of manual labor, our handiwork. All have a direct bearing on the development of science in the West.

Meaningful History
Here I shall present the case that in order for modern science to arise, time and history had to be understood as meaningful, purposeful and progressive. Contrary to popular opinion, I shall argue that the Greeks did not lack a concept of history as progression any more than they lacked a concept of time as succession. It was the *quality* of time and history, however, that was wanting to the Greek mind. Rather than conceiving of time and history presented as an opportunity for humanity to improve their lot, the Greeks saw time and history as murky, chaotic reflections of an original harmonious state of perfection. Time was understood as an enemy, a destroyer. It was like the grim reaper, the white-cloaked, macabre angel of death in the medieval death dances, who swept down all before him with indiscriminate strokes of his mighty scythe.

European philosophical and theological thought has, to a large extent, accepted Augustine's argument that the Greeks regarded time as cyclical. In contrast, the Judeo-Christian concept of time was linear. As Augustine put it, the Greeks thought that "the motions of the sun, the moon and the stars constituted time". Christians, however, believed that there is "a present of things past, a present of things present and a present of things future".[53] Augustine's explanation, popular though it is, does not stand up to critical analysis. It is thus disconcerting to realize that several generations of theological students, have been taught Peter Brunner's thesis that contrasts Augustine's understanding of time with the Greek understanding of history.

Oscar Cullmann's argument in *Christ and Time* picked up and popularized the idea so that it is often taken for granted that the Bible's linear understanding of time stands over against the Greek "cyclical conception".[54] Investigation into the basic sources, however demonstrate that Thorlief Boman is quite correct in maintaining that Cullmann's thesis cannot be sustained.[55] As Boman points out, Cullmann's citation of Aristotle's statement in this regard, "for indeed time is some kind of circle", is inconclusive because it is taken out of its original context.[56] Aristotle does indeed state, "There is some kind of circle of time." He ends the paragraph, however, with the statement: "The whole measurement of a thing is naught else but a defined number of units of its measurements." This indicates that Aristotle means to stress the *measure* of time and not its *circular nature*.[57]

It is true, as Augustine argues, that the concept of the circularity of all things was natural for the Greeks because they saw both rest and movement in the circular orbits of the heavenly bodies. Augustine is also accurate in pointing out that the Greeks analyzed time in spatial modes. These were best understood as segments of a circle that was likened to the orbits of the heavenly bodies. These orbits, for the Greeks, were continuous, repetitious and eternal. Thus it is not difficult to understand why Aristotle's concept of time, explained as it is in analogy to orbital motion of the heavenly bodies, has often been taken to indicate that time itself is circular and cyclical.

More basic to Aristotle, however, is the notion that time "like motion is a perpetual succession". This is best represented as successive motion along a line or as a series of segments of a line. Thus Aristotle sums up one part of his argument by saying, "We conclude, then, that time is the numeration of continuous movement without any qualification, not only of some particular kind".[58] The fact that near the beginning of the argument he states, "A partial revolution is time just as much as a whole one is",[59] should have warned scholars like Brunner and Cullmann that, although Aristotle uses a circle as the illustration of the line along which time proceeded, his interest was primarily enumeration, not circularity.

Aristotle's representation of the continuous successive movement of time by the supposed circular orbit of the sun as it describes its way across the heavens reflects his concept of the geocentric universe. Even today although we have accepted the heliocentric nature of the planetary system, we continue to think of time as being measured by the sun's orbit. Hence we still speak of the times of day as "sunrise", "noon-time", and "sunset". For Aristotle, however, the time-line *per se* might have been either circular or straight. Further, the measurement of time as movement along even a circular line rules out neither succession nor sequence. To repeat, such measurement does not mean that time as such is circular or repetitious even though the model by which time is illustrated, i.e., the sun's orbit, was thought to be circular. For Aristotle it is the eternality of the motion of the heavenly bodies that, having neither beginning nor end, commends them as "time-keepers". As he put it, time will never end, "for it is always at a beginning".[60]

Peter Brunner's attempt to contrast Augustine's "Christian concepts" of time and eternity with that of the Greeks suffers from his not having investigated the Greek concepts thoroughly enough. It is true, as Brunner points out, that the Greeks longed for that pure, enduring, "eternal now" without yesterday and tomorrow. The desire of the Greeks was "the totally filled beautiful moment which rested and remained in itself and swallowed both past and future into itself".[61] However, Brunner's attempt to contrast Augustine with the Greeks rests on the thesis that for Augustine "the brilliance of eternity which remains always at rest" belongs to God alone.⁶² His argument falls flat especially in comparison to the teaching of Plato and Aristotle. In that Plato's whole philosophy was based on the differentiation between perfect eternity and corruptible time, the rest of "the same" and the movement of "the other" (in Aristotle "the same" is reflected in the unmoved mover, "the other" in the heavens which were moved), Augustine comes closer to following the Greeks than to disagreeing with them.

For his part Plato in the *Timaeus* speaks of the difference between the eternal living being of the "father and creator" and the "creature which he had made". It is in contrast to the eternal father that "the moving and living creature was the created glory of the eternal gods".

> Wherefore he [the father and creator] resolved to have an image of eternity, which he made when he set in order the heaven moving according to number, while eternity rested in unity; and this image we call time.⁶³

This "image", called "time" by Plato, which is the whole of a life, including its time span, and which is progressive and is measured off in numbers as a succession of units, is clearly differentiated from eternity which is complete and at rest. In the *Timaeus* Plato's distinction between time and eternity is such that eternity, as differentiated from time, *is not* endless astronomical time. Time, by contrast, *is* endless astronomical time. "Eternity" is the life form to which God (the Father and Creator) belongs. It stands in contrast to the form of the world which God created as an "image of eternity". The latter is characterized as "moving according to number", whereas eternity is "rested in unity". Days, nights, months and years are all parts of time. The past and the future are "created species of time".⁶⁴ Hence, to quote Boman in reference to Plato, "Time is only a pictorial, moving imitation of immovable and unalterable eternity which represents perfection."⁶⁵ Time is of creation. It "is as unbounded as the world and just as finite".⁶⁶ Whereas eternity is everlasting, the world is "growing and decaying, waxing and waning".⁶⁷

Aristotle, in his *Physics*, comes to the same kind of conclusion as does his master, Plato. The relationship of predecessor and successor τὸ πρότερον καὶ τὸ ὕστερον is an analogy from space. "The time occupied is conceived as proportionate to the distance moved over."[68] Thus also "the primary significance of before-and-afterness is the local one of 'in front of' and 'behind'".[69] Since there is a "before and after", in space or "in magnitude" as Aristotle put it, "there must be a before and after in movement in analogy with them". This movement then is the seat of before-and-afterness in respect of both "movement and time". Aristotle makes this explicit when he said, "[it is] when we are aware of the measuring of movement by a prior and a posterior limit that we may say that time has passed".[70]

Hence, rather than regard the Greek concept of time as being circular and simply as being repetitious so that, as Cullmann would have it, life is purposeless, a simple coming around, the Greeks think of time as a dimension of succession by which events are recognized as being past, present or future. Whether it is measured by units on a circular or a rectilinear line, as Boman points out, is epistemologically of no importance.[71] What is important, in contrast to the later teleological notion of time as conceived of in the Judaeo-Christian tradition, is the quality of reality as affected by time and this is the whole difference. For Aristotle, time is negative.

> We do not say that we have learnt [anything] or that anything is made new and beautiful, by the mere lapse of time; for we regard time in itself as destroying rather than producing, for what is counted in time is movement, and movement dislodges whatever it affects from its present state.[72]

Hence, all reality which is subject to time, *i.e.*, history itself, runs downhill. It suffers degradation and is in a state of decay.

Following the pattern depicted by Hesiod in the eighth century B.C., life degenerates from perfection to imperfection. Since the advent of history humankind has descended through five races. From the time of the golden and blessed race when human beings were gods and the earth produced rich harvests, the process has been one of devolution through the silver race, the copper race, the race of heroes and finally the iron race. In contrast, as Aristotle understands reality, those things which are eternal such as eternal "forms" and geometric figures remain forever intact.

> Things which exist eternally, as such, are not in time; for they are not embraced by time, nor is their duration measured by time. This is indicated by their not suffering anything under the action of time as though they were within its scope.[73]

Yet, time, is of history and history is characterized by chronological movement in a definite direction. This movement is depreciative rather than appreciative. It progresses but instead of inclining upward or keeping on one level, it declines and moves toward decay. Moreover, this downward movement is not the result of the human condition as such. Rather it is due to the direct intervention of the gods at critical points in history. Again to quote Boman:

> Hesiod speaks of no history and no development; Cronos developed the entire golden race from which after its earthly life protecting spirits arose for later races. Subsequently an entire new race, the silver race, was created by Zeus, which lived its life and then died out. The reason for the silver race's not being of the same quality as its predecessor is to be found not in human guilt but in divine will. The next generations were also created anew by the gods. The descriptions of the five generations are five independent pictures devoid of internal relation; nor does Hesiod speak of an uninterrupted descent of mankind, for between the copper and the iron races, Zeus created the glorious race of Heroes.[74]

Thus rather than regard the Greek view of history as circular and recurrent, as do Brunner and Cullmann, for instance, it would seem more accurate to speak of it as a time in which degeneration takes place. History is a largely devolutionary process characterized by suffering, misery, deterioration and chaos. Actual time is antithetic both to the harmonious, complete and restful image of eternity and to the golden age of the past. It stands in sharp contrast to the imagined far-away places such as Elysium, the Isles of the Blessed, Atlantis and the Gardens of Alcinous where the conditions once enjoyed by the golden race continue, to a certain extent, to exist.

It is at this point that we see the relevance of revolutionary understanding of time and history in the Judaeo-Christian faith. The Genesis story like Greek mythology informs us that creation was originally harmonious and perfect. The world created by God is good. It is of utmost importance, however, to note that in the biblical tradition, not even "the Fall", which signifies the entrance of evil into the world, changes the fact that creation is restorable to goodness.[75] Neither in the Old Testament nor in the New Testament does the Fall provide any excuse for humanity to err, stray from the ways of God or avoid seeking perfection (Mt. 5:28). Rather, for Israel restoration, which includes reformation, is both a promise and a command. Repentance, turning from evil toward good, indicates the proper direction of life. God who is "the first and the last" (Is. 44:6, 48:12, Rev. 1:17) has the whole of history under control. In the end there will be a new heaven and a new earth (Is. 66:22, Rev. 22:1). Hope is in the future. When the kingdom toward which God moves history eventually arrives, Israel and even its former surviving enemies will go up to Jerusalem year after year "to worship the King, the Lord of hosts, and to keep the feast of booths" (Zech. 14:16).

In Christian theology the concept of *Heilsgeschichte* (salvation history) continues this scheme. It is true that salvation history sometimes has differentiated too sharply between the history of "the people of God" and the rest of history. Thus in a gnostic kind of way, the fall was considered only a moral problem. People were thought to have to be saved out of the world rather than with it, because the world was considered irredeemably evil. The impact of the New Testament as a whole, however, is to emphasize the Old Testament conception of the solidarity between humanity and the rest of creation, and to look forward to a coterminous salvation. Even fallen creation is said eventually to be restored and as restorable in the present. In this way time and history are transformed from a nihilistic regression into the netherworld of nothingness into the hope-filled possibility of restoration. History is understood as the possibility of movement toward the fulfillment of the hoped for perfect creation which is to come.

Karl Barth's theology developed a unique approach to this New Testament insight. He stresses in his concept of the covenant that not only has the world itself been chosen by God as the stage on which he will work out the history of salvation, but the whole of creation is caught up in the movement of reconciliation and restoration.[76] There is hope for a new heaven and a new earth and even now creation "waits with eager longing for the revealing of the children of God . . . because the creation itself will be set free from its bondage to decay and obtain the glorious liberty of the children of God" (Rom. 8:19 ff.). Thus, in dependence on God's providence, time itself becomes pregnant with positive possibilities. History, as measured by time, is the "space" wherein humanity exists in partnership with God working with and within creation for the restoration, the improvement, the salvation of the whole of creation. As Butterfield notes:

> The transition from the Old Testament to the New, and the notion of a Kingdom of the Father, succeeded by a Kingdom of the Son, with a Kingdom of the Spirit to follow, were examples of this. It has been suggested that the modern idea of progress owes something to the fact that Christianity had provided a meaning for history and a grand purpose to which the whole of creation moved.[77]

Thus, the life of humanity and nature is seen to have a dynamic, a flow and a purpose which not only make life worth living but give an ultimate dimensionality to all finite activity. Life is the arena of restoration, and humanity is meaningfully involved. The active, saving God both reconciles humanity and institutes the ministry of reconciliation (2 Cor. 5:18). As Barth has put it, "The time between the Ascension and the Parousia should not simply be empty . . . it should not remain without a provisionary fulfillment of the promise . . . that the kingdoms of this world will be the kingdom of God and of His Christ".[78] In Calvin's terms, the doctrine of divine election had assured one's salvation. One could thus safely dedicate oneself to work in the world using the gifts one has been given not primarily for the self but for the benefit of all.[79] The Lord is the Lord of this life as well as the next. Hence life is purposeful. Those faithfully engaged in fulfilling life's promises may rejoice in life and enjoy its gifts here as well as hereafter.[80]

Following S. Jaki, this is the root of "The Leaven of Confidence" stemming from the Christian faith. It is a *sine qua non* of the kind of science and technology that developed under the teleological orientation of the culture of the West which has been informed by the Christian faith.[81]

Labor's Rewards

The three Judaeo-Christian concepts, (1) that nature could be considered dependable and restorable to the good, (2) that humanity was called into a purposeful relationship with God in the redemptive process, and (3) that history had meaning, influenced directly the removal of the ancient Hellenistic stigma against manual labour. The Greeks considered intellectual endeavours to be the occupation of gentlemen while work with one's hands was considered to be the employment of underlings and slaves. In stark contrast, because the Judaeo-Christian faith looked forward to the restoration of humanity and nature, it encouraged not only mental labor but manual labor as well.

The rise of science is inconceivable without the appreciation of the value of work as encouraged by the Judaeo-Christian faith. Consequently, we must devote some effort to understanding the Greek concept of manual labor and their abhorrence of it. The Greek duality between the realm of the spiritual and the realm of the material accompanied a fundamental mind-body dualism. The "spiritual" was considered to be real and important and the material was understood to be shadowy and secondary. Consequently the mind was esteemed while manual labor was disdained. The pervasive dichotomy between mental and manual labor was reflected in Greek social structure as a matter of course. The social structure, in turn, reinforced the basis for this dichotomous conceptuality. Even today in the countries of the Levant, manual labor is so scorned that men who might otherwise be taken as "laborers", taxi drivers, office boys, clerks or what-have-you, often sport grotesquely elongated fingernails on the little finger of one or both hands as a badge of their non-manual, hence non-menial occupations.

The achievements of Greek architecture, the utilization of plumbing, the use of the water wheel with geared transmissions for power, the manufacture of weaponry, all indicate what for their time was a highly sophisticated technical ability.[82] Archimedes' (287-212 B.C.) description of the five basic machines of mechanics — the lever, the wedge, the wheel, the pulley and the screw — and Hero's air and water powered machines reinforce this point. Nevertheless, technique and thought, practice and theory, like manual labor and mental labor, were neither seen as integrally necessary to one another, nor were they brought together in a co-ordinated whole. Max Born makes the point when he writes in reference to the Greek physicists: Thales, Anaximander, Anaximenes, Leucippus (fl. 5th cent. B.C.) and Democritus (c.460 - c.357 B.C.):

> These Greek gentlemen lived in a world which venerated the harmony of beauty of body and mind. They despised manual work which was the task of slaves, and thus they neglected experiment which cannot be done without soiling one's hands. Thus, no empirical foundation of the ideas was attempted, nor their technical application, which might have saved the antique world from the assault of the barbarians.[83]

Hence, although Archimedes, himself a mathematician and physicist, invented fantastic contrivances for the defence of Syracuse, we have no evidence that he wrote anything at all about the machines. G. Howe is probably right when he says of Archimedes, "As a great theoretician, only his research into mechanical principles was worthy of being noted down. This neglect, however, has caused his technical achievement to be forgotten."[84] Such an attitude runs hand-in-glove with Plutarch's (c.46 - c.120) report that Plato became angry over the fact that some of the important mathematicians of his circle had begun to devote themselves to "mechanics". As Georg Picht (1913-1982) records, if

Archimedes was right in his report about Plato, Plato went so far as actually to banish the study of mechanics from mathematics and from pure science altogether.[85]

The distrust of mechanics and technology in the philosophical schools of Greece is nicely illustrated by the myth of the two Titan brothers, Prometheus and Epimetheus, as told by Plato through the mouth of Protagoras the Sophist. At the creation Zeus had appointed Prometheus and Epimetheus to equip his creatures, both animals and humankind, with the necessities of life. Epimetheus took on the task of distribution leaving to Prometheus the duty of inspection. Epimetheus was singularly successful at giving the animals the qualities for self-preservation — swiftness, the ability to burrow, strength and armaments, etc. When it came to humankind, however, he had nothing left. When Prometheus made his inspection, he found that humans stood naked, barefooted and, most significantly, weaponless. Hence, they were unable to protect themselves from the wild beasts. At this juncture Prometheus, seeing his brother's perplexity, came to the rescue by stealing both the mechanical arts of Hephaestus and Athene and the fire necessary for their utilization.

Plato makes a point of noting that the fire which Prometheus stole was not the kitchen fire of Hestia. Rather, Prometheus stole both fire and the technical intelligence which accompanied it. The fire was the fire of the forge. It was the fire for making tools which were necessary for the arts of Hephaestus and Athene, i.e., technology. It was the fire necessary for the production of the weapons of warfare. As it turned out, however, technology and the art of weaponry were as much of a curse as a blessing. They were ambiguous at best. Technology and fire were not gifts of the gods. Rather, Prometheus had taken the well-guarded arts from the common workshop of Hephaestus and Athene by stealth. In stealing the arts along with fire, Prometheus had provided humankind with the technical means to move beyond their designated and proper limits. Humanity had thus seized power not proper to its nature.

Little wonder, then, that as the tale goes on we are told

that even with the power of technology and with the weapons resulting from its application, humanity was incapable of coping with their difficulties. Alone and without government "of which the art of war was a part", the individuals were no match for the wild beasts, even when they were armed. If one attempted to manage by moving into cities, this technical ability did more harm than good. Instead of using weapons to protect themselves from the predatory beasts, they turned on one another. Only the actions of the divine powers finally restored order. Zeus himself intervened and sent Hermes to teach the populus the political arts by which they were enabled to live together in community.[86]

Thus, as over against Protagoras the Sophist (whom Plato portrays as appreciating reality only at a superficial level and who argues for the necessity of technique as the means to overcome people's impotence) Plato makes the point that technology is essentially foreign to humankind. At best it is a necessary evil. Ultimately, however, it is disastrously dangerous and can be kept in check only when people equipped with it are put under the supervision of government.

Although Plato may well have a point in showing the effects of technology gone awry, his basic condemnation of technology is indicative of the philosophical attitude that prevented it from possibly becoming the handmaiden of Greek physics. Greek mechanics thus remained an auxiliary to military science. As long as this type of mind-set ruled there was little chance that science, as we know it, and for which both manual labor and technology are absolutely essential, could come into being.

By contrast, according to biblical understanding, manual labor belongs to humankind as a proper function and its results are commended. It is true that, according to the creation saga, part of the consequence of Adam's having been driven out of paradise was that *in toil* he should eat from the cursed ground and *in the sweat of his face* he should eat bread (Gen. 3:17 ff.). Hence, work is work. However, the fact that the author of Genesis 1 pictures God as having labored at creation and the writer of Genesis 2 shows humankind in paradise *before the Fall* dressing and keeping the Garden, indicates that, for the Hebrew mind, sin is not the primary reason for labor. Work is not considered a curse as such. The fallen state of creation makes labor less rewarding than was the case prior to the Fall, but work is not condemned. In Genesis 9 Noah became a "tiller of the soil". This recognizes and emphasizes that to work with one's hands is to follow God's command. It is a proper occupation of one whom God has called into his covenant (Gen. 9:20). It may also be worth recalling that the Hebrew *abad,* which is translated, "to till", also means "to serve", whether with one's hands or to serve God in worship.

In contrast to the Greeks, then, for the ancient Hebrews manual labor was considered an honorable occupation. The heavens are considered the "work" of God's fingers and the earth the "work" of his hands (Ps. 8). Hard work, whether that of a craftsman (Pr. 22:29) or the wife (Pr. 31:10 ff.), is praised. Labor is blessed (Ex. 34:21, Ps. 127:1, Pr. 10:5, 21:5, Ec. 5:12, *et al.*) and honest labor has its reward (Gen. 26:12, 29:10 ff., Ps. 107:36-38, Ec. 2:24, *et al.*). Idleness, however, is a curse (Ec. 10:18). Even as slackness is condemned (Pr. 10:4) so the ambitious ant stands as an example against lethargy (Pr. 6:6 ff.) In that God had labored six days in creating the world before he rested on the seventh day (Gen. 2:2), labor was regarded as much a part of the covenant command as was respect for the Sabbath, "six days shalt thou labour" (Ex. 20:9-11).

This same attitude toward the worth of work is expressed within the New Testament accounts. Jesus, the handworker (carpenter—Mk. 6:3) is reported as saying, "My Father is working still, and I am working" (Jn. 5:17). Some of Jesus' immediate disciples were fishermen — Simon, Andrew, James and John (Mk. 1:16-20) — when he called them to be his followers. The Apostle Paul who was a tentmaker (Acts 18:3) charged the Thessalonians to work with their hands (1 Th. 4:11). The author of Second Thessalonians condemns those "living in idleness ... not doing any work". He reiterates the command, "If anyone will not work, let him not eat", and exhorts the idle "in the Lord Jesus Christ to do their work in quietness and to earn their own living" (2 Th. 3:10 ff.). The very strength of the New Testament exhortations, however, indicates the difficulty of the problem in the Graeco-Roman culture into which the Christian faith spread.

The development of the appreciation of manual labor within Western culture has had significant consequences. Although the Benedictine discipline of *ora et labora,* pray and work, emphasizes honoring God on one's knees rather than in working with one's hands, both worship and work are stressed as being important before God. So, too, Brother Lawrence (c.1605-1691) showed that a monk could wash the pots and pans to the glory of God as well as honor him by prayer and praise. With the Reformation which considered not only ministers but all members of the church to be servants of the Lord and ministers of reconciliation, the work of the non-ordained was rated of equal worth with that of the ordained. Martin Luther could speak of the work of a mother in the nursery, of the milkmaid in the barn and of the farmer in his field as being as much a part of the worship of God as were the prayers of a monk. John Calvin's (1509-1564) emphasis on industry was such that Max Weber (1864-1920) can make him partly responsible for the modern capitalistic movement. To quote Calvin himself, "No task will be so sordid and base, provided you obey your calling in it, that it will not shine and be reckoned very precious in God's sight".[88]

Hence, when the opportunity arose, the use of one's hands in making instruments and the use of instruments to examine the properties of nature, whether that of the macrocosm with the telescope or the microcosm with the microscope, could be considered an honorable occupation. Working with one's hands was thought to be as appropriate a posture for the Christian as hands folded in prayer. Mental labor and manual labor became equally honorable and equally necessary activities. Hands used to make instruments were as basic to the rise of the new science as were the hands that manipulated them in observation and experimentation. We know that Tycho Brahe was active in fabricating most of the instruments by which he was able to fix the positions of the planets and stars. Butterfield can say that it is proper to picture Galileo:

> ... passing his time in a sort of workshop with trained mechanics as his assistants, for ever making things—even making things for sale—and carrying out experiments, so that in him the mechanic or artisan combined with the philosopher to produce a modern type of scientist.[89]

Newton spent untold hours fashioning prisms and the lenses for his reflecting telescope. With the re-evaluation of human activity, especially after the Reformation, we also see a new attitude to "secular learning". The study of philosophy and scientific treatises could be rated on a par with the study of the Bible or theological works.

More and more Aristotle began to be studied as preliminary to the study of medicine as well as basic to the study of theology (as was the case, by the way, already in the early sixteenth century at the University of Padua). Thus, it is no coincidence that Butterfield can cite Bernard de Fontenelle (1657-1757), the secretary of the French Academy of Science (1699-1741) as including in his biographies of scientists student after student who was intended for the church. Having been put off by the prevailing educational methods or because of being given "an education in mere words and not in real things", they became scientists instead.[90] When Pierre-Sylvain Régis (1632- 1707) came across Cartesian philosophy, he was immediately struck by it. Louis Carré (1663-1711) found that Cartesianism opened up a new universe for him. Pierre Varignon (1654-1722) picked up a volume of Euclid by chance and moved from theology to the study of Descartes. After four years of study for the priesthood, Jacques Ozanam (1640-1717) left his clerical studies "out of piety and love for mathematics".[91] Thus, Butterfield notes that the biographies read, "like Christians recounting conversions in the early stages of a religious movement, when one man after another sees the light and changes the course of his whole life".[92] We could add Kepler to the list. Kepler quit the study of theology for the study of mathematics, but was convinced that he was continuing to do theology. He studied the ways of the heavens not only to the honor of God but to trace his ways in the courses of the stars.

As said above, it would be wrong to picture the ancient world and that of the Greeks in particular as having no technology. They had developed it to a considerable degree. Nonetheless the fact that they lacked the mind-set that allowed them to respect technology was certainly a factor which prevented the development of empirical science in Greece beyond a rather rudimentary level. Technology could not become the essential handmaiden of science. Thus nothing provided the tools necessary for the extension of scientific knowledge. For this reason the Greeks were severely handicapped.

If the analysis above is correct, then it is legitimate to argue that science and western civilization have been expedited both by the divorce between the sacred and the secular, and athe marriage between mental and manuel labor. It is also true that the biblical teachings re-emphasized by the Reformation, such as the priesthood of all believers and the importance of the dedicated Christian life, added much-needed fuel to the fire. Thus, although there is no belying the fact that a great deal of the basics necessary for the development of science came to the West from the Greeks, Greek learning alone did not prove to be a sufficient fundament for modern science. Valuable as were the teachings of ancient Greece with regard to (1) the observation of nature, (2) the application of mathematical measurements to these observations and (3) the development of a largely coherent and ordered system of knowledge, these contributions must be considered as having been preliminary to the development of experimental science rather than as adequate for its foundation. Modern science did not develop in Greece, not only because "it hadn't the chance to come that far", as is often said. Rather, it did not arise because there were elements within the thought of both Plato and Aristotle which stood in direct antithesis to the attitudes and concepts that were necessary for the development of modern science.

Butterfield, as we have said, has long since noted the

"ecclesiastical flavor" of the teaching of Aristotle and his successors. We have noted the debilitating effects of such concepts as the hierarchical structure of the heavens, the perfect circularity of the movements of revolving spheres, the necessity of intelligences to move the planets, the grading of the elements of nature in the order of their nobility, and the view that celestial bodies were composed of an "incorruptible fifth essence". These, along with Aristotle's concept of *deus sive natura* and his notions of history and labor, made it highly improbable that empirical science would ever have developed on the basis of Aristotle's thought. As Butterfield puts it, "Indeed, we may say that it was Aristotle rather than Ptolemy who had to be overthrown in the sixteenth century."[93] Aristotle had to be put aside before experimental science could really begin.

In sum, although we owe the beginnings of science to the Greeks, we also owe to them concepts and influences that retarded the development of science. It was, therefore, only through a process of both selection and recasting of concepts of Greek learning, in what Howe refers to as "the crucibles of a thousand years of Christian influence and history in western Europe", that science became the "life ruling power" which we know it to be today.[94]

It is hardly a coincidence, therefore, that it was just at those points where the Aristotelian system is called fundamentally into question that the groundwork both for experimental science and for the Protestant Reformation was laid. The biblical doctrines stressed by the Reformation encouraged the continuation of the process.[95] In their re-emphasis on the sovereignty of God in creation and the freedom of nature as contingent on God alone, the Reformers specifically denied first, Augustinian sacramentalism in relationship to the world; second, the Aristotelian idea of the interpenetration of God in nature; and third, the exclusivity of Aristotle's logic. Consequently, the Reformed doctrines of history and salvation gave new meaning to life and human activity.

I am not arguing that Reformation thought is completely adequate. Although both Luther and Calvin moved against

Aristotelian theology and the neo-Platonic sacramentalism of Augustine with reference to the world, they too continued to exhibit a heavy dependence upon Platonic thought. This dependence is exhibited especially through the influence of Augustine's "neo-Platonism". This is particularly true for Luther, for whom "eternal realities" continue to take precedence over earthly matters. His *Zwei Reichen Lehre* (doctrine of the two realms), for example, continued the legacy of Ockham in dividing reality between the sacred and the secular. Thus, Friedrich Nietzsche's (1844-1900) characterization of Christianity, especially German Lutheran Protestantism, as "Platonism for the people", may be justified. Yet in Calvinistic thought, there is a greater appreciation for God's works and human responsibility in relationship to life on earth. His advice that to know nature it was necessary "to observe and analyze the works of creation", as Hooykaas reminds us, indicates the direction the development of science was to take.[96]

It is the Reformers' insistence on seeing the world through the "spectacles of Scripture", as Calvin put it, which was most important for opening the eyes and the minds of those who became responsible for science. The emphasis upon biblical doctrine over against that of hellenized theology meant that there was a renewal of the importance of recognizing God as Creator and Lord over the whole of creation. Creation could thus be seen as creation under God's lordship but freed from any divine interpenetration. Humanity was responsible for nature. This, along with the notion that time was valuable and history was hopeful, led to the belief that work was meaningful. Therewith a new vision of the world arose. The result was that the mind-set of western Europe was transformed. When Greek science and learning became known, curiosity about nature, combined with the empirical habit of mind and inductive logic worked out especially by Duns Scotus and William of Ockham, caused the explosion of scientific enthusiasm that we rightly refer to as the scientific revolution.

Thus, with the Reformation the monolith, which had structured western understanding of the relation of faith and science and had been inspired by Aristotelian metaphysics and Thomistic theology, collapsed. For the Reformers God was the God of truth from whom truth came. Revelation as based upon Scripture became the sole source of theological truth; and the world of nature became the sole source of scientific truth. This refocusing of both theological and scientific conceptualities along with the concurrent breakdown of the medieval political structures allowed scientific and technological activity to be understood as "God-given".

When coupled with the Reformation emphasis on the positive value of creation, history and manual labor, this reassessment of science not only allowed science the freedom it needed for its development but encouraged it to be an ally in the search for knowledge about God's good creation. It is more than coincidence, therefore, that with the victory over Aristotelian metaphysics, on the one hand, and the Reformation-inspired breaking of the political hammerlock of the medieval church over Europe, on the other, that science moved from the south to the north. It moved from the lands of the Renaissance in the south of Europe and developed most vigorously in those of the north that were under the influence of Protestantism.[97]

In the process the whole power structure and face of European culture changed. In the late Middle Ages at the height of Arab learning and influence the Mediterranean, in Butterfield's words, for all intents and purposes became "a Mohammedan lake". The Renaissance was located largely in southern Europe. After the Reformation the locus of new learning, which was itself being reshaped by science and technology, moved northward to the lands bordering the English channel. Here, to refer again to Butterfield, both England and Holland held leading positions. France's role in promoting the new order was played to an astounding degree by the Huguenots. After the Edict of Nantes in 1685 and their ensuing exile from France, they became the "productive nomads of Europe". Those who fled to England shuttled scientific ideas back to their own country via journals written in French and published in Holland. Those

188 The Renaissance, the Reformation and the Rise of Science

who moved into the Ruhr area of Germany became responsible for the first technological and industrial development of that country.[98]

In England the Puritans were largely responsible for the promotion of both science and industry. This indicates that there is a definite relationship between the influence of Calvinist thought and scientific and technological development. The one outstanding exception was the Dutch Calvinist Gisbert Voetius (1589-1676) who was recaptivated by Aristotelian rationalism and scriptural literalism and specifically forbade the study of Descartes. G. Howe, whose evidence is corroborated by Reijer Hooykaas, reports that the Puritans literally dominated the Royal Society. In 1663, a year after it was chartered, forty-two of a total of sixty-eight members were Puritan.[99]

When the Calvinist theological tradition became transposed into the political thought of John Locke (1632-1704) democratic political structures began to replace aristocratic ones, then political and scientific forces blended to form the basis of modern industrial society. In the year 1876 at Philadelphia's World's Fair commemorating the one-hundredth birthday of the United States, Europeans could observe for the first time that just one hundred years after James Watt (1736-1819) fabricated the first industrially viable steam engine, American industry was on its way to surpassing the achievements of European technology.[100] This development may well undergird Max Weber's thesis that the capitalistic economic system which, for good or ill, is unthinkable without western scientific technological development, received much of its mind-set from the Reformed theology of John Calvin.[101]

The scientific and technological movement of the industrial revolution had became a world force. Whether in the Old World or the New, the movement was northern European and British. As Butterfield has said,

'It is probably correct to say that Protestantism had become the ally . . . of a new form of civilization. The center of gravity of the globe itself seemed to have changed and new areas of its surface found for a time their 'place in the sun'.[102]

The result was literally the reshaping of both our natural and our human environments.

THE MOVE TOWARD NATURALSCIENCE

Francis Bacon's England
Within the argument of this work I have brought forward ample evidence for the contribution of Christian scholars to the birth of science. In doing so I have demonstrated that although there was an uneasy relationship of early Christian natural philosophers with the ecclesiastical authorities, they considered themselves faithful sons of the Church of Rome. Such scholars as Roger Bacon and William of Ockham prior to the Reformation, the master, Robert Grosseteste, and his fellow Franciscan Duns Scotus, as well as the more esoteric neo-Platonic and Hermetic thinkers, Marsilio Ficino and Giordano Bruno, were devout believers in the Catholic faith. Yet all were to a greater or lesser extent anti-Aristotelian and thus anti-Thomist; and all considered the rigidity of the Thomist deductive methodology, the basis for the medieval synthesis, to be far too narrow for "natural philosophy". All believed that there was a methodological gap that must be closed if reality was to be disclosed. The natural science that was struggling to be born required a new method and doctrine if it was to have a future.

Francis Bacon responded to this apparent methodological gap. He stands on the edge of the Renaissance, which included the Protestant Reformation, and the scientific revolution of the seventeenth century. By tempering the esoteric,

Platonism and Hemeticism, with the scriptural doctrines emphasized by his Puritan contemporaries, Bacon became the midwife of a new epoch in scientific thought. Francis Bacon perhaps more than any other brought the then fragile child of science to birth. A short excursus into Bacon's biography will help us gain perspective on this accomplishment.

Bacon's biographer, James Spedding (1808-1881) sets the scene of the background of Bacon's life. Only a year before his birth, the country's confession of faith had been quite suddenly changed from Catholic to Protestant. England became "the stronghold and refuge of the Protestant cause in Europe."[103] Consequently England was in turmoil during the reign of Elizabeth who had ascended the throne in 1558. Elizabeth's father, Henry VIII, effectively broke with the Church of Rome by passing the Act of Supremacy (1543) declaring that the King was the only supreme head of the Church of England. His son, Edward VI, consolidated Protestantism by the adoption of the Book of Prayer produced by Thomas Cramner and the Forty-nine Articles. Mary I, who during the reign of her half-brother Edward defied the Act of Unification by celebrating Mass in her private chapel, reversed the decisions of Henry and Edward. She re-established the authority of the Pope and brought back Roman Catholic doctrine and practice. Then Elizabeth I restored the independence of the Church of England. Upon doing so she was faced with Roman Catholic attempts at restoration. Rebellion and counter-reformation agitated by Roman Catholic missionaries sent from the continent threatened the peace of her rule. In addition the challenge of the Puritans compounded her problem. While in exile on the continent under Mary I they had imbibed the Calvinistic understanding of doctrine and polity and hence objected to the established church as vehemently and, as it turned out, more effectively than the Roman Catholics. This crisis continued unabated. The restoration of the Prayer Book of 1552, the reduction of the Forty-two articles to the more Catholic Thirty-nine (the Prayer Book and the Forty-two Articles had been introduced under Edward VI) along with the maintenance of the episcopacy, the establishment of Catholic ritual and the Act of Supremacy (1559) that restored ecclesiastical supremacy to the crown satisfied neither Calvinist nor Roman Catholic.

It was in this historical context that Bacon lived his life. After his studies at Cambridge (1573 and 1575) and a sojourn in France (1579) he began to study law at Gray's Inn in London. In 1582 he became a "barrister", or lecturer at the Inn, and later a "bencher", or senior member. Subsequently Bacon entered "Her Majesty's" civil service. Like his father before him, he was to spend most of his active adult life serving the British crown.[104]

He became Counsel Extraordinary, first to Queen Elizabeth (1533-1603) and later to King James I. James raised him to the rank of Solicitor General, then to Attorney General, and finally to Clerk of the Star Chamber. He was knighted and became Lord Keeper of the Seal, and lastly, Lord Chancellor and Viscount St. Albans.

It was a boon to science although a personal disgrace that in 1621, while Lord Chancellor, Bacon was convicted of bribery, fined, and sent to the Tower for a short period. The pomp and circumstance of his career in civil service went by the board. After his release Bacon devoted himself until his death in 1626 to his writings, especially to those propounding his new method of the interpretation and understanding of nature. Like Einstein who claimed to have his first flash of insight concerning relativity at age 16, Bacon began to consider the necessity for a *Novum Organum* as early as age 15 while still at Cambridge. It was not until he was 46, in 1607, that he settled on the plan for his *Great Instauration*.[105]

Bacon followed his father who was Keeper of the Seal under Elizabeth in service to the crown. Yet it was from his mother, who was "a learned, eloquent, and religious woman, full of affection and puritanical fervor", deeply interested in the condition of the Church and a devoted non-conformist, that he inherited his fervor whether directed to serving the state or the cause of science.[106] His disposition was to serve him well during these very troubled and complex times.

troubled and complex times.

Upon Mary I's (1516-1558) death Elizabeth I succeeded to the throne. Protestantism was again re-established through the Acts of Conformity and Unconformity (1559). Protestant exiles, who in exile on the continent had learned Puritan ways were free to return. Roman Catholics attempted to regain power first by the rebellion of the Earls of Northumberland and Westmoreland (1569), and then by agitation from Jesuit missionaries from the continent. Reaction led to a law declaring it treason for a Roman Catholic priest to set foot on English soil (1585). Likewise, a law against conventicles (1593) drove underground Puritans who, along with the Roman Catholics, objected to the established church..

Although Bacon was Puritan in both doctrine and temperament, he remained a loyal subject to the crown. He saw both the Roman Catholic attempts at the restoration of the Church of Rome and the Puritan attempts to disestablish the established church as threats to the stability of the nation. Although his father Nicholas was a mediating figure between Protestants and Roman Catholics, [107] he is known to have supported the non-conformists (those Puritans who refused to follow the Act of Uniformity of 1559).[108] This same sentiment is reflected in Francis Bacon's argument against liturgical uniformity. He stood for unity rather than uniformity.[109]

As Gary Deason has pointed out with reference to Bacon's later scientific thinking in which he was careful to distinguish things divine from the things of nature, Bacon strongly opposed some of the innovations that the Genevan-influenced Puritans had violently sought to impose upon the church.[110] In Deason's words, "they were so infatuated with the Kingdom of God that they ignored the welfare of the Kingdom of Man."[111] They were thus in danger of defacing "the laws of charity and of human society."[112] Although Bacon's sympathies were Puritan, he opposed the radicals who fought to force the Presbyterian form of government upon the established church. The point of contention centered on the purist's belief that Scripture authorized only one form of church government.[113] Bacon re-

The Reformation and the Rise of Science

of "but one form of discipline in all the churches." Rather

> God had left the like liberty to the church government, as he hath to civil government, to be varied according to time and place and accidents which nevertheless his high and divine providence doth order and dispose.[114]

After spending two years in Paris Bacon entered Parliament. In Paris he experienced at first hand the chaos caused by intrigue. He was afraid that this might be exported to England.[115] His fears were compounded when William of Orange (1533-1584) was killed by an assassin's bullet in 1584. In the face of attempts on Elizabeth's life, Bacon wrote a letter to the Queen in which he advised against draconian measures against the Roman Catholics, and urged moderation. They might be schooled into becoming Protestants, he believed.[116] Certain that the truth would make its own converts, he urged that diligent preachers be appointed to each parish, so that "careful catechizing and diligent preaching would bring about a diminishing of the papist number."[117] Bacon knew full well that the disunity of Puritans only strengthened the hand of the Roman Catholics. In fact the "religious dissentions" were, as he said later, Spain's only hope for a successful invasion of England in its attempt to restore the Church of Rome on the island.[118] Hence Bacon hoped to restore unity to dissenting parties by a campaign of enlightenment.

On the one hand, Bacon stood by the Puritans. He did not favor disestablishment of "those who have sought to deface the bishops."[119] Like his mother who testified that she had "profited more in the inward knowledge of God his holy will" from Puritan thought than she had ever gleaned from "hearing odd sermons at Paul's,"[120] Bacon also praised the Puritans for their biblical preaching. He was convinced that "the best of those observations upon texts of Scriptures that had been made in sermons by the Puritans represented the best work of diversity since apostolic times."[121] He regrets that England's universities did not train ministers "to preach soundly and handle the Scriptures with wisdom and judgment."[122] On the other hand, he accused the Puritan radicals of attempting to prevail over the establishment "with uproar and violence". The result was "faction and division".[123] Although Bacon was critical of the bishops and recognized that they had failed to eliminate abuses, by and large he respected them as men of "great virtues." This narrowness revealed a fatal flaw in Puritan thought. They desired to eliminate every "non-scriptural" practice in church and society without realizing that Scripture like all else "hath its limits."[124] In his view the Puritans were wearing uncritical blinders. Unable to compromise, they were unable to see the good in the established church.

Bacon feared that the chaos in France could spread.[125] He recognized that religion was often the cause of "perturbations, seditions, and wars",[126] and advised that changes in government "should follow the example of time itself" which although it "innovateth greately" does so "quietly and by degrees scarce to be perceived."[127]

For Bacon the Christian faith was not a program for radical social change. "*Ira viri non operatur justitiam Dei*, the wrath of man worketh not through the righteousness of God."[128] Much less was it above the law after the motto, "away with the law and it by force", which called the people to sedition and mutiny.[129] He could not abide the hubris of the ultrazealous Puritans who had "impropered to themselves the names of zealous, sincere and reformed, as if all others were cold minglers of holy things and profane and friends of abuses."[130]

Rather than reputation by *ictus inermium* (headless arrows), Bacon commends good works as the Christian cause. He states that scripture teaches people to respond to the non-religious according to their works. He refers to the second table (the last five of the Ten Commandments),"*a man doth namely boast of loving God whom he hath not seen, if he love not his brother whom he hath seen.*"[131] So much for those who "fancying themselves worthy of a closer conversation with God neglect the duties of charity toward their neighbour, as inferior matters."[132] Hence Christ's miracles were merciful and helped to relieve people's distress.[133]

Indeed Bacon was intent on reform. He intended his Instauration to be a reform of learning. Learning, however, was to be judged by its capacity for good works and human benefit. "In religion we are warned that faith be shown by works". Therefore, "it is altogether right to apply the same test to philosophy. If it be barren, let it be set at nought".[134] There is no learning simply for the sake of learning. Rather, "Knowledge is to be limited by religion."[135] More particularly, since it is "a charter from God", it "must be subject to the use that God hath guaranteed, which is the benefit and relief of the state and society of man."[136] Thus, "knowledge bloweth up, but charity buildeth up."[137]

In a real sense, Bacon was a pragmatist. At the age of thirty-one he sought a position at court so as to have the means to confront "those who are guilty of frivolous disputations, confrontations and verbosities", and those who spend their time at "blind experiments and auricular traditions and impostors." Also abstraction and speculation have "too much enhalted and glorified, to the infinite detriment of man's estate."[138] In their place Bacon hoped to offer "industrious observations, grounded conclusions, and profitable inventions and discoveries". Good thoughts are "little better than good dreams, except they be put into act."[139]

Accordingly Bacon's Instauration of philosophy was to contain "nothing empty or abstract." It was to be a total turning away from Aristotelian philosophy which "wove spider's webs wonderful for their texture, but useless for any practical purpose."[140] Bacon's renewal of learning was to be "designed to improve the conditions of human life".[141] The goal was to "establish a chaste and lawful marriage between Mind and Nature," a marriage that would bring forth "wholesome and useful inventions to war against our human necessities and, so far as may be, to bring relief therefrom".[142]

The secret of the matter lay in nature itself. Bacon was convinced that "the discovery of the new arts, endowments and commodities" were of the greatest benefit "for the bettering of man's life." If a man could succeed in kindling a light in nature, it "should presently disclose and bring into sight all that is most hidden and secret in the world . . . that man (I thought) would be the benefactor of the human that man (I thought) would be the benefactor of the human race . . . the propagator of man's empire over the universe, the champion of liberty, the conqueror and subduer of necessities."[143]

Central to Bacon's scheme of renewal was the role of the "magus". At this point his biblically-influenced Calvinism and his inherited Renaissance understanding of humankind as the priest of nature conjoined.

> Believing that I was born for the service of mankind, and regarding the care of the commonwealth as a kind of common property which, like the air and water belongs to everybody, I set myself to consider in what way mankind might best be served and what service I was best fitted to perform.[144]

From this we learn that there was another side of the Elizabethan mind that Bacon inherited. This was the Renaissance romanticism, such as had been illustrated in the New Science of Ficino. As S. L. Bethell points out this romanticism can be considered "the common property of Europe in the sixteenth century".[145] Here we meet again the hierarchical structure that we have seen over and over again in neo-Platonism and its successors which connects earth to heaven and all things in between. This "vast chain of being"[146] continued to dominate the western mind from the Renaissance until, as Arthur Lovejoy has pointed out, it finally became the horizontal chain of Darwinian evolution.[147]

The Protestant "Scientist" Francis Bacon and The Hermetic Imagination

Few figures in the history of the development of science are fraught with the mystery surrounding Francis Bacon. At first glance it might seem as if his adherence to "the ancient neo-Platonic, neo-Pythagorean Hermetic-like occult philosophy" is antithetical to his "scientific method". As we shall see, however, this is not the case. Rather, it was Bacon, a child of both the Renaissance and the Protestant Reformation, who demythologized the magical science of the Renaissance and brought its concepts down to earth by straining its concepts through the sieve of biblical doctrine.[148]

As far as Bacon's general intellectual milieu is concerned, Bacon admits having read Patrizzi and Telesius as well as Anaximenes, Anaxagoras, Democritus and Parmenides. He did not study Pythagoras, however, judging him to be superstitious.[149] Rather, he compares himself to Demosthenes (c.384-322 B.C.), Cicero (c.102-43 B.C.), and Seneca (c.4 B.C.-A.D.65).[150] The three, like him, were banned and devoted themselves to writing.[151] He also spoke of the philosophies of Aristotle, Plato, Democritus, and Hippocrates. The latter were highly vigorous "at first, [but] by time degenerated and imbased".[152]

However, since "truth is contrary" and "time is like a river which carrieth things which are light and blown up, and sinketh and drowneth that which is sad and weighty",[153] the opinions of Democritus were "passed by" while those of Plato and Aristotle, because they were "both agreeable to popular sense, and the one was uttered with subtlety and the spirit of contradiction, and the other with a style of ornament and majesty, did hold out".[154] It is no doubt Plato and Aristotle whom Bacon has in mind when he speaks of philosophy and the intellectual sciences standing "like statues, worshipped and celebrated, but not moved or advanced".[155]

As we have seen, this "anti-Aristotelianism" is not at all new with Bacon, but as far as the development of science is concerned is as old as Philoponos. In Bacon, however, it seems not to have been prompted by this tradition, but by Henry Cornelius Agrippa von Nettesheim (1486-1535). In his work on the vanity of the Arts and Sciences von Nettesheim had attacked the learning of his day for its pedantry and subservience to Aristotle. Bacon wholeheartedly shared and reiterated this opinion.[156] Yet, while Bacon's alternative to Aristotle was the eventual development of a method of induction and a logic of discovery, von Nettesheim, like Cusa before him and Bruno and Robert Fludd (1574-1637) afterwards, was content to remain committed to an Hermetic type of neo-Platonism and the magic of the Cabbala.[157]

With regard to Hermeticism *per se,* it is probable that Bacon did know of Giordano Bruno. The very year that Bacon was finishing his studies in law at Grey's Inn Bruno sought refuge in England and lectured in London. D. W. Singer points out that Bacon's lyric references to infinity in his *Novum Organum* resemble lines in Bruno's *The Ash Wednesday Supper*. Further, Bacon's suggestion of spiral rather than circular motion of the heavenly bodies may be due to Bruno's influence.[158] Yet as Singer also indicates, in Bacon's introduction to his *Historia naturalis et experimentalis* (1622), he refers to Bruno as someone who depends upon imagination rather than experiment for knowledge.[159] Here he rejects the spiral motion of planets along with "dragons" as the stuff of "dreams" and "fancies".[160]

Bacon's acquaintance with William Gilbert's (1544-1603) work on magnetism, also suggests that he must also have known of his Hermetic tendencies.[161] In addition Bacon is also acquainted with Robert Fludd, the English Hermeticist par excellence.[162] Bacon's comparison of James I (1566-1625) to *Hermete illo Trismegisto* leaves little doubt that he was more than somewhat enamored by the "ancient theology".

> There is met in your Majesty a rare conjunction as well of divine and sacred literature as of profane and human; so as your Majesty standeth invested of that triplicity which in great veneration was ascribed to the ancient Hermes, the power and fortune of a King, the knowledge and illumination of a Priest, and the learning and universality of a Philosopher.[163]

Bacon also repeats the Hermetic tradition that "Moses . . . is said to have been seen in all Egyptian learning, which nation was early and leading in matter of knowledge".[164] Perhaps even more indicative of an Hermetic influence is that along with the name "Valerius Terminus", which Bacon gives to the author of *Of the Interpretation of Nature*, he names the commentator, "Hermes Stella". This commentator is supposed to shed star-like light upon the work. Unfortunately nothing remains of the annotations attributed to "Hermes Stella", if indeed Bacon ever got around to writing them.[165]

We have already mentioned Bacon's familiarity with the writers of the encyclopaedic tradition of the Renaissance.

His dependence on Bernardino Telesio (1509-1588), who constructs his universe on a combination of passive matter, earth, and the active principles of heat and cold, is obvious. Bacon simply seems to accept that heat elevates and rarefies bodies and produces white. The cold is associated with the earth; it solidifies, prevents motions, and produces black. Consequently these concepts become a part of his *History of the Dense and the Rare* and his *Novum Organon*.[166] This is witnessed to again when Chardan's story of the winds and sacrifices on Mount Olympus and Athos, where the tops of the mountains are so high that "there had been neither wind nor rain" so the ashes lay undisturbed for a full year, is repeated in Bacon's *Historia Ventorium (History of the Winds)*.[167] Furthermore in his *New Atlantis*, his work portraying a Utopian land beyond the pillars of Hercules, the evidence continues to mount. One reads of Solomon's Temple, a College at the Sixth Day and a spirit of brotherly love.[168] Certainly Bacon's reference here to Persian, Chaldean and Arabian men as bearers of religious, occult and Hermetic science indicates a knowledge of Hermetic materials. In addition he singles out Tyre, the city of Pythagoras, where the King Hiram built the Temple.[169]

Thus, although the whole effort of Bacon's emphasis on science and the advancement of learning was concentrated on the attempt to find a "better method" than Aristotle's "barren study of nature", his search was far from purely or integrally scientific. He points out that the appearance of a new star in Cassiopeia signifies the heavens' protest against Aristotle's cardinal doctrine of the eternity, perfection and unchangeableness of the heavens.[170] Even his own ideas were far from free of the attraction of the ancient, the mystic, and the occult. For him it is a legitimate use of *poesy* and the *parabolic* "when the secrets and mysteries of religion, policy or philosophy are involved in fables or parables.[171] He admits his intention of:

> Discoursing scornfully the philosophy of the greacians [*i.e.*, especially Aristotle]wth some better respect to ye Aegiptians, Persians, Caldes and the utmost antiquity and the mysteries of the posts.[172]

To this end he wrote *De Sapientia Veterum (The Wisdom of the Ancients)* specifically to interpret these fables as representing "treasures recovered from antiquity" to prepare men's minds to receive his new knowledge as presented in the *Instauration.*[173] He introduced these ancient fables of Greek mythology in which he found "true virtue", "genuine prophecy", and full depth with a verse of his own:

> Rich mine of Art: Minion of Mercury;
> True Truch-man of the mind of Mystery:
> Inventions storehouse; Nymph of Helicon:
> Deep Moralist of Time tradition:
> Unto this Paragon of Brutus race
> Present thy service, and with cheerful grace
> Say (if Pythagoras believ'd may be)
> The soul of ancient wisdome lives in thee.[174]

A table then follows with thirty-one characters of Greek mythology from "Cassandra or Divination", to "The Sirens or Pleasure", each of which Bacon treats in separate short chapters explaining the wisdom each had to offer to humankind. It is especially in Chapter 12, "Coelum or Beginnings", Chapter 13, "Proteus or Matter", and Chapter 28, "Sphinx or Science" that Bacon showed the relevance of what he considered to be the gist of ancient wisdom deemed relevant to the seventeenth-century "scientific mind".

"Coelum or Beginnings" is based on the fable of Coelum (the heavens), the most ancient of the gods. In this story Coelum is rendered impotent by Saturn who emasculates him. Saturn himself produces many children, but has the murderous proclivity of devouring his own offspring. Of all the children Jupiter alone escapes. He then turns on Saturn and pares off his genitals and throws them into the see, and Saturn into Hell. Venus is born from the sea and, as Jupiter now reigns, he is forced to defend his kingdom in two wars. In the first he is invaded by the Titans, but with the help of Sol, the only Titan favoring him, Jupiter survives. In the second battle, he is invaded by the giants, but after routing them with thunderbolts, he reigns secure. The fable which, according to Bacon, differs from the opinions of the philosophers, outlines the beginning of the world. It is these opinions "which Democritus afterward laboured to maintain, attributing eternity to the first Matter and not to the world", coming close, thereby, to "Divine Writ".[175]

"Proteus or Matter" was Neptune's herdsman, a seer and a prophet "that he might well be termed thrice excellent" (a reference no doubt to Hermes Trismegistus).[176] Proteus lived in a cave and those desiring advice from him shackled him in manacles to hold him fast. Even manacled, Proteus like nature, was not totally subject to his captors. Rather, following the canons of alchemy, even manacled he remained free to turn himself into the forms of fire, water, and beast, and thereby to return to his own form.[177] This story illustrates Bacon's idea that it was necessary to bend nature and force it to give up its secrets. He explained that the fable seems to unfold the secrets of nature and the proportions of matter. He Christianizes the fable, however, by comparing it to creation when the divine word produced matter at the creator's command.[178]

For Bacon, the fable of the "Sphinx or Science" is of paramount importance for pointing to science. In that his science also included applied science, or in our terms "technology", Bacon valued the parable, for, as he said, it "contains within it no less wisdom than elegancy . . . especially that which is joined to practice".[179]

> It is diverse in shape and figure by reason of the infinite variety of subjects wherein it is conversant. A maiden face and voice is attributed unto it for its gracious countenance and volubility of tongue. Wings are added because Sciences and their inventions, do pass and flee from one another, as it were in a moment, seeing the communication of Science is as the kindling of one light at another. Elegantly also it is feigned to have sharp and hooked talons, because the Axioms and arguments of Science do so fasten upon the mind, and so strongly apprehend and hold it, as that it cannot stir or evade, which is noted also by the divine Philosopher. Ecc. 12.11 *Verba sapientium* (saith he) *sunt tanquam aculei, et veluti clavi in altum defixi.* The words of the wise are like goads, and like nails driven far in.[180]

Like the Sphinx, science was placed in steep and high mountains. Its position was among the "lofty and high things, looking down upon ignorance with a scornful eye".[181]

> Furthermore, Science may well be feigned to beset the high ways, because which way soever we turn in this progress and pilgrimage of humane life, we meet with some matter or occasion offered for contemplation.[182]

The two kinds of riddles associated with the Sphinx, some concerning "the nature of things", others "the nature of man", refer to the "two kinds of Emperies as rewards to those that resolve them". Thus they define the scope of science as an interaction between humankind and nature.[183]

> For the proper and chief end of true Natural Philosophy is to command and sway over natural beings, as bodies, medicines, Mechanical works, and infinite other things.[184]

Thus, although Bacon was not himself actually a scientist his continuation of the Renaissance tradition of interpreting "ancient wisdom" as a cryptic apology for science led him to be prepared to understand the whole of reality in "scientific terms". From the beginning nature was "shot through" with scientific rationality. That rationality had become sidetracked, in a way not unlike that in which the ancient wisdom and the pre-Socratics became sidetracked by Aristotle's rationalistic deductive system. Now was the time for it to again come to the fore. Bacon intended his *Novum Organum, Advancement of Learning,* and *Natural History* to announce and promote the new age of science. In contrast to his predecessor Roger Bacon, and to his contemporary Johannes Kepler, who like Roger Bacon, was influenced by the Hermetic tradition, he was not convinced by the neo-Platonic insistence on the importance of numbers and of mathematics.

Bacon's scientific disposition was incomplete on two counts. First he was influenced by Hermeticism and enamored by ancient myths and fables, and second, he lacked like Bruno an appreciation of the use of mathematics in science. Nevertheless, or perhaps just because of the power of this romantic vision of the mystery and harmony of all things, he took the material world seriously. Consequently more than anyone else of the seventeenth century, Bacon was responsible for translating Ockham's "empirical habit of mind" into an *apologia* for experimental science.

Faith, Nature and Scientific Method

He also carried on this tradition in another way. As with his predecessors from Philoponos to William of Ockham, Francis Bacon's concepts of the Christian faith play no small part in his efforts to promote natural science. We find that his works abound in biblical quotations and references.[185] In the "Prooemium" or "Preface" to his *De Interpretatione Naturae* (1603), he speaks as a Calvinist. He thought of himself as a person destined for service, called as Ficino felt himself called, albeit from a quite different source.

> Believing that I was born for the service of mankind, and regarding the care of the commonwealth as a kind of common property which like the air and water belongs to everybody, I set myself to consider in what way mankind might best be served, and what service I was myself best filled to perform. Now among all the benefits that could be conferred upon mankind, I found none so great as the discovery of new arts, endowments and commodities for the bettering of man's life.[186]

Bacon acknowledged his dependence on the "goodness of God" for the progress of his work. Although "luck" and "accident" as well as the "felicity" of the times, rather than "wit", play major roles.[187] He knew better than to transfer the ways of sense gained "by natural light" to theology. As "things human are not to interfere with things divine", so the senses discover natural things but "darken and shut up the divine".[188] Thus, although there is no basis for a "natural theology", there is no contradiction between knowledge of nature and knowledge of the divine. Neither does the one replace the other. On the contrary:

> From the opening of the ways of sense and the increase of natural light there may arise in our minds no incredulity or darkness with regard to divine mysteries, but rather that the understanding being thereby purified and purged of fancies and vanity, and yet not the less subject and entirely submissive to the divine oracles, may give to the faith that which is faith's.[189]

Contrary to Aristotelian philosophy and the Thomistic theology it inspired, the knowledge of the world is, for Bacon, not an avenue to the knowledge of God. For, if any think that "by view and enquiry into sensible and material things to attain to any light for the revealing of the nature or will of God, he shall dangerously abuse himself".[190] Knowledge of the world "must be subject to that for which God hath granted it; which is the benefit and relief of the state and society of man".[191] Consequently, Bacon recommends "the work of the inventor". Bacon's inventor's achievements are felt everywhere and forever "not in some particular invention but in kindling "a light in nature". This light, strangely, or not so strangely, is similar to Bruno's light. In his *La cena de le ceneri* he describes it as:

> a light which should in its very rising touch and illuminate all the border regions that confine upon the circle of our present knowledge; and so spreading further and further should presently disclose and bring into sight all that is most hidden and secret in the world.[192]

Then, recognizing in himself that, as he says, "I was fitted for nothing so well as the study of the Truth", Bacon proceeds to lay out his intention for "the interpretation of nature", which was to take place "in steps of ascent until a certain stage of Generals be reached".[193] Hence, it is no coincidence, as Torrance has pointed out, that Bacon's understanding of faith allowed him to realize the opportunities and duties not only in "his service of God in the Kingdom of God (*regnum Dei*) but in the building up of his own life on earth through inductive science in the kingdom of man (*regnum hominis*)".[194]

As with Roger Bacon, so with Francis Bacon, Christian faith encourages the increase of natural knowledge both because it "leadeth to greater exaltation of the glory of God", and because it preserves "against unbelief and error". Such knowledge, in fact, restores man to the sovereignty and power which he had when he called the creatures by their true names at the beginning of creation.[195] Man is created to understand nature. "God hath framed the mind of man as a glass capable of the image of the universal world, joying to receive the signature thereto", and also to "discern those ordinances and decrees which throughout all these changes are infallibly observed".[196] Not that man is capable of understanding all things, for God has reserved "the highest generality of motion or summary law of nature" within "his own curtain". Nevertheless, the operations "within man's sounding" are both "many and noble".[197]

The object of the study of nature is the knowledge of nature. Its purpose is to move beyond knowledge and to benefit life. The "working and discursion of the spirits" in the brain attained "by the sweat of the brows" is the "amplification of the power and the kingdom of mankind over the world".[198] The object is to "let the human race recover that right over nature which belongs to it by divine bequest, and let power be given it; the exercise thereof will be governed by sound reason and true religion".[199]

Bacon warns that, as in "the inquiry of divine truth", men have "ever inclined to leave the oracles of God's word", so in the "inquisition of nature they have ever left the oracles of God's works", and depended instead on "the deceiving and deformed imagery which the unequal mirrors of their own minds have represented unto them".[200] He thus reiterates the point of Roger Bacon and William of Ockham that the object of scientific investigation is nature itself. He ends the first chapter of *Of the Interpretation of Nature* appropriately by warning that, as it is true of the kingdom of Heaven, so "in the kingdom of knowledge", no one shall enter "*except he become first as a little child*".[201]

This sums up Bacon's new disposition which originated while he was a young student at Trinity College, Cambridge, where he became disenchanted with what he learned. At this early point in his life Bacon was disappointed with professors,"who neither knew or aspired to know more than was to be learned from Aristotle".[202] Even while the new star appeared and then disappeared in Cassiopeia, thus "protesting by signs and wonders against the cardinal doctrine [the eternal perfection of the heavens] of Aristotelian philosophy", his mentors continued to be bound by Aristotelian scholasticism. For Bacon, this disposition had all the fecundity of a desert.[203] At age 15, he wrote, "If our study of nature be thus barren . . . our method of study must be wrong: might not a better method be found?"[204] By this point in his life his new disposition had come to full flower.

Four centuries earlier Roger Bacon had complained about an epistemology whose foundation was deduction and syllogism. At this point Francis Bacon added to this complaint the observation that Aristotle remained the "dictator" of schoolmen who "out of no great quantity of matter, and infinite agitation of wit, spin out unto us those laborious webs of learning which are extent in their books".[205] "To go beyond Aristotle by the light of Aristotle is to think that a borrowed light can increase the original light from whom it was taken".[206]

As Kuno Fischer argues, Francis Bacon began to question Aristotle not on the basis of science as such, but from the point of view of Renaissance philosophy, one of whose founders was Telesius and of which Bruno and later Fludd were representatives.[207] It is this philosophy of nature, according to Fischer _ '*de rerum natura juxta propa principia*' which is "the fruits of the resurrection of antiquity (*Altertum*)" — that unfolded during the sixteenth century in Italy. It formed the transition and last link between the boundary (*Grenze*) of the middle and the newer philosophy.[208] Bacon was born into an age on the move. The pillars of Hercules had been conquered. The transatlantic voyages had opened the new world for investigation.[209] For Bacon, the vision of the *New Atlantis*, that land of Utopian science, was about to be discovered. To quote Fischer again, where he differentiates the Renaissance attitude from that of Augustine, "*Das 'regnum hominis' tritt an die Stelle der 'civitas Dei*'" (The 'City of God' was replaced by the 'Kingdom of man.').[210] The new world was discovered. It was time to renew the human race.[211]

Bacon sees the relationship between the pre-Socratics, Parmenides and Democritus, and of the Renaissance Telesius.[212] It is the pre-Socratic atomist Democritus whom Bacon[213] looks to as the precursor of the basic method he elucidates in his *Instauratio Magna*.[214] It is true, of course, that Bacon criticized Democritus for failing to develop experimentation.²¹⁵ Yet as R. L. Ellis (1817-1859) points out, this led him to respect Democritus for regarding matter as eternal and inseparable from form, tracing the essence of things to ultimate particles.²¹⁶ In this way he could affirm the fact that Democritus offers an ancient alternative to Aristotle, while he looked for an alternative elsewhere.

Bacon's scientific disposition is perhaps best spelled out in "The Epistle Dedicatory to King James", the preface to his *Great Instauration.* He first claims that his ideas are "quite new, totally new in their very kind." Yet, "they are copied," he says, "from a very ancient model." The model consists of three aspects: "the world itself", "the nature of things", and "the mind".²¹⁷ In intention, Bacon is neither a realist who thinks ideas are a simple reflection of reality, nor is he an idealist who thinks that reality is wholly dependent upon categories of mind. Rather, he affirms Democritus' argument that, "The truth of nature lieth hid in certain deep minds and caves".²¹⁸ The apprehension of reality is dependent on the mental process that probes into the deepest recesses of nature. Thus, after expressing "suspicion of things long established", and stating that it was time for "the regeneration and restoration of the sciences",²¹⁹ Bacon pleads with the King to collect and perfect "a Natural and Experimental History". Only then might "philosophy and the sciences no longer float in air, but rest on the solid foundation of experience".²²⁰ "Experience," for Francis Bacon, as indeed for Roger Bacon, is to be understood as "experiment".

As we noted earlier, Francis Bacon in the "Preface" to the *Instauration,* criticised the philosophy and the intellectual sciences of his time for becoming "statues" which are "worshipped and celebrated but not moved or advanced".²²¹ "Time is like a river, which has brought down to us things light and puffed up, while those which are weighty and solid have sunk."²²² He adds, this sad state of affairs has come about because none have built "upon experience and the facts of nature as long as is necessary".²²³ The universe is not plain to see, rather it is like a "labyrinth" which is "knotted and entangled".²²⁴ It is simply not the nature of nature to give up its secrets easily or by observation alone. Rather, nature like Proteus, who would reveal his identity only when bound, gives up its secrets only when handcuffed.[225]

Hence, nature cannot be known by the "natural face of man's judgement," by "accidental felicity", "excellence of wit", or "chance experiments". Rather, what is needed is a "method" — steps guided by a clue of the senses "laid out on a sure plan".[226] Although the "facts of nature", he argues, are far from obvious, ultimately they are beyond discovery, at least to the nth degree. No one can perfectly investigate "the nature of anything in the thing-itself". Although he may "vary his experiments" at will, "he never comes to a resting-place but still finds something to seek beyond".[227]

Nature is to be encountered with "the true and legitimate humiliation of the human spirit". Only thus is progress possible that might allow the "images and rays of natural objects [to] meet in a point as they do in natural vision". It is "things in themselves" and the "concordance of things" that are to be seen. This is "the true and lawful marriage between the empirical and the rational faculty" which is to be established.[228] The process is pursued in "obedience to the everlasting love of truth".[229]

Like Ockham Bacon is interested in things and the reality of things. He is not an experimentalist who sells books and builds furnaces thereby "forsaking Minerva and the Muses as barren virgins", and relying instead on Vulcan. He does not intend to abandon reflective meditation.[230] On the contrary, knowledge demands the "perpetual working and exercise of the mind".[231] Thus, the "mental operation" that simply "follows sense" is rejected. Instead, Bacon proposes to "lay out a new and certain path for the mind to proceed in, starting directly from simple sensuous perception". Thereby the mind will be instructed from the outset.[232] The result is to be "certain and demonstrable knowledge", not "pretty and probably conjectures".[233]

> The End of our Foundation is the knowledge of causes, and the secret motions of things (*et motuum, ac virtutum interiorum in Natura*) and the enlarging of the bounds of Human Empire, to the effecting of all things possible.[234]

Francis Bacon's Way of Science

In his *Opus Majus* Roger Bacon cited four hindrances to truth: the use of insufficient authority, custom, popular opinions and the concealment of ignorance which hides behind a display of apparent knowledge. Similarly Francis Bacon designated four kinds of "idols" that "beset men's minds" of which the understanding must be purged if the way of science is to proceed. The first, the *idola tribi (idols of the tribe)*, are shared by humanity in general and forms a kind of common sense. They mislead the mind by causing those under their sway to ascribe to nature an over-abundance of order and regularity. They promote the prejudiced selection of instances for observation so that those supporting preconceptions are included, and those opposed thereto are left out. Generalizations are made from far too few instances and the senses are weakened. Consequently rather than evidence being garnered from observation, conjectures are put forward. The second, the *idola specus (idols of the cave)*, are designated as "individual commonsense" based upon one's particular mind-set, whereby all things take on the hue of one's favorite subject. This includes reverence for either the ancient or the modern as well as an over-evaluation of either the differences or resemblances among things. The third, the *idola fori (idols of the market place)*, are vulgar public opinions influenced by popular argument and causing words to posit non-existent entities. Thus, names are given to generalities based on observation of an insufficient number of objects. The fourth, the *idola theatri (idols of the theater)*, are erroneous opinions garnered from false philosophies and false demonstrations.[235] The false use of logic, unwarranted authority, ordinary common sense, both communal and individual, public opinion and false philosophizing are all set aside. In addition, the superstitions that all celestial bodies move in circles, spirals and dragons, the fiery orb above the air and the ten to one density ratios of the elements are all rejected as fancies.[236]

These are, of course, the negatives of what Bacon thinks

a proper scientific stance ought to include. A proper scientific stance is to open the mind to the world. "For the world is not to be narrowed till it will go into the understanding (which has been done hitherto), but the understanding to be expanded and opened till it can take in the image of the world, as it is in fact."[237] In order to accomplish this Bacon prescribes "a league between these two faculties, the experimental and the rational", which enables the world to be "altered and digested" by the understanding.[238] The wisdom principally derived from the Greeks, which is but "the boyhood of knowledge" and has the characteristic property of boys, "can talk but it cannot generate; for it is fruitful of controversies but barren of works".[239] Former discoveries which "lay near to the senses and immediately beneath common notions" were made "by practice, meditation, observation, and argumentation. However, to reach the hidden parts of nature "it is necessary that a more perfect use and application of the human mind and intellect be introduced".[240]

The end of it all, as stated above and as evident in the emphasis of the theology of the covenant, is humanity's reattainment of its proper place as sovereign over nature. Once again "he shall be able to call the creatures by their true names and shall again command them", regaining the power "which he had in his first state of creation".[241] The end, which Bacon realizes cannot be given by the method itself, is all important for "the end rules the method".[242] The procedure is to set forth "everything relating both to bodies and virtues in nature". Insofar as possible these are to be "weighed, measured and defined".[243]

The purpose, to emphasize it again, is not speculative. Rather the goal is "practical working" which arises from "the due combination of physics and mathematics".[244] The proposed result is a natural history with five major divisions: (1) questions, (2) descriptions of experimental processes, (3) things doubtful or questionable, (4) account of miscellaneous observations, and (5) review and specification.[245] In his "History" Bacon sets out "a plan of work" and includes a catalogue of the planned particular histories which are to

The Reformation and the Rise of Science

incorporate its results. On this basis future science will proceed. The catalogue lists 130 specifics concerning "the heavenly bodies", "masses", "species", "man", "his anatomy", "characteristics", "diseases", "occupations", "arts", "products", and "mathematics". However, as Bacon explains, mathematics consists of "observations rather than experiments".[246] He thereby expresses his ambiguity toward mathematics.

In our interpretation we have been following the recommendation of A. E. Taylor and Adolfo Levi to study first Bacon's conception of the scope of science, its model of classification and general scientific view of the world, and then to move on to the *Novum Organum* and its theory of generalizations.[247] Accordingly Bacon's method is found largely in Book II of the *Novum Organum*.[248] In Book I Bacon sets out his scientific *apologia*. Here he attacks Aristotle, "who made his natural philosophy a mere bond-servant of his logic, thereby rendering it contentious and well nigh useless",[249] and whose "inquisition of Final Causes is barren, and like a virgin consecrated to God produces nothing".[250] In contradistinction Bacon intends to replace Aristotle's *Organum* with his *Novum Organum*. His method, Bacon claims, sets out "means which have never yet been tried".[251] He overlooks, thereby, or perhaps did not know of Grosseteste and Roger Bacon. However, as V. F. Whitaker contends, Bacon may in fact, be dependent on the French philosopher Petrus Ramus (1515-72), who in attempting to move away from the Aristotelian and scholastic traditions used induction and a method of arrangement from the most to the least conspicuous and to whom Bacon makes reference.[252] At any rate, Bacon was convinced that "the sciences we now possess are merely systems for the nice ordering and setting forth of things already invented. Hence, he can even go so far as to comment that Aristotle's "wisdom and integrity" are shown in his "diligent and exquisite history of living creatures".[253] What was needed, however, were "methods of invention or directions for new work" (*modi inveniendi* and *designationes novorum operum*).[254]

The plan was to derive axioms from "the senses and particulars, rising by a gradual and unbroken ascent" until arriving "at the most general axioms last of all".[255] Like Duns Scotus and Ockham, he intends to discover the reality of nature by concentrating on the things of nature themselves. Hence, after commenting on thinkers ancient and modern, and saying that he observed great differences in point of faculty (*facultatis*), and in point of truth (*veritatis*), Bacon goes on to say, "The sum of the matter is this — if men will submit themselves to things, something will be done; if not those wits will come round again in the circle (*redibunt in orbem*)".[256]

> Everything is dependent upon never allowing the eyes of the mind stray from concentration on the things themselves so it is enabled to receive the images clearly. But God forbid that we should give out a vision of our own imagining for a pattern of the world.[257]

Like Duns Scotus and Ockham, Bacon proposes a method of induction by which axioms are to be "duly and orderly formed from particulars" which "easily discover the way to new particulars, and thus render the sciences active".[258] In contrast to the "*anticipations of nature*" Bacon called this an "*interpretation of nature*".[259] Rather than being arrived at by deduction, this "reason" which is "elicited from facts by a just and methodological process"[260] leads to experimentation so that nature which "to be commanded must be obeyed"[261] may reveal itself by itself "working within".[262] Thus, deduction is set aside; and induction, along with experimentation, which Bacon is certain is new in the history of human knowledge, is proposed as the foundation of his *Novum Organum*. This Bacon also calls, "true directions concerning the interpretation of nature".[263] The result is a "natural history" which is to take in nothing except instances, connections, observations and canons. Bacon boasts, "from my method of philosophizing (*ex nostra philosophandi methodo*) there will be gathered by the way an abundant crop of useful works; of which speculations and disputations yield few or none".[264] Thus, in an aphorism which sums up much of what we are really beginning to appreciate especially since Albert Einstein (1879-1955), Bacon states:

It is idle to expect any great advancement in science from the superinducing and engrafting of new things upon old. We must begin anew from the very foundations, unless we would revolve for ever in a circle with mean and contemptible progress.[265]

Bacon sets out his new method most distinctly in *Novum Organum*, Book II where the depth of his insight can be seen in the first aphorism. Rather than understanding nature on its surface Bacon states that the work and aim of "human power" is "to generate and superinduce a new nature of natures" on a given body. The work of human knowledge is to discover "the form", or the "true specific difference", the "source of emanation" or, perhaps best, the *natura naturans* (the nature-engendering nature) of a given nature. In addition, Bacon proposes to discover the "transformation of concrete bodies" and the process which is incurred "from the manifest material to the form which is engendered."[266]

Like Ockham, Bacon is convinced that in nature "nothing really exists beside individual bodies". These, however, perform individual acts "according to a fixed law". It is "this very law, and the investigation, discovery, and explanation of it", which is the foundation of knowledge.[267] "Knowledge", for Bacon, is "a discovery of all operations and possibilities of operations from immortality (if it were possible) to the meanest mechanical practice".[268]

In order to avoid the term "causes" as used by Aristotle whose "efficient and material causes", though of some use, do not "touch the deeper boundaries of things", Bacon prefers the term, "forms". These both embrace "the unity of nature in substances the most unlike" and allow "truth in speculation" as well as "freedom in operation".[269] Bacon compares a "form" to a "rule of operation" which is *certain, free, and disposing or leading to action*.[270] He describes it in words similar to those which Ockham uses with reference to "cause". One wonders if more than coincidence may be present here.

> For the Form of a nature is such, that given the Form nature infallibly follows. Therefore it is always present when the nature is present, and universally implies it, and is constantly inherent in it. Again, the Form

is such, that if it be taken away the nature infallibly vanishes. Therefore it is always absent when the nature is absent, and implies its absence, and inheres in nothing else. Lastly, the true Form is such that it deduces the given nature from some source of being which is inherent in more natures, and which is better known in the natural order of things than the Form itself.[271]

As in Roger Bacon, Duns Scotus, and Ockham, the unified nature of nature is taken for granted. "True forms" are universal.

Bacon, then, proceeds to set down procedures by which individual or "simple natures" such as gold can be identified and investigated. There is a "latent process" in the generation or transformation of bodies which though "continuous" for the most part "escapes the senses". Only by careful observation and analysis by comparing body with body, and thus noting what escapes and what remains, what is continued and what cut off, what propels and what hinders, etc., can the "latent process" be ascertained. Since "every natural action depends on things infinitely small, or at least too small to strike the senses",[272] the direction of investigations is from the obvious and particular to the more and more minute and general. Thus, the *"latent configuration"* which bodies possess may be sought by "reasoning and true induction with experiments to aid". The method aims at "a reduction to simple natures and their Forms, which meet and mix in the compound".[273]

Such activity transforms "the complicated to the simple, the incommensurable to the commensurable", the absurd to rational quantities, and "the infinite and vague to the finite and certain". Indeed, although as indicated, Bacon was in no sense competent in mathematics, such inquiries into nature as described "have the best result when they begin with physics and end in mathematics".[274] "Physics", for Bacon then, has to do with investigations into "efficient cause", "matter", "latent process", and "latent configuration". Metaphysics has to do with what, "in the eye of reason at least, and in their essential law", are "eternal and immutable Forms".[275] To physics is added the practical division of "mechanics"; and to metaphysics, as if to show Hermeticism had not been entirely forgotten, is added "magic". Magic is added "on account of the broadness of the ways it moves in, and its greater command over nature".[276]

Thus, the "two generic divisions" which are absolutely essential to experimental science and for the interpretation of nature are "induction" and "deduction". Important, however, is that induction is first. It is "to induce and form axioms from experience", and requires the use of "sense", "memory", and "reason" (what we would call "intuition"). Second, it is to deduce and derive *experiments* (not answers or substances as such) from axioms. For, as Bacon himself points out:

> ... the understanding, if left to itself and its own spontaneous movements, is incompetent and unfit to form axioms, unless it be directed and guarded. Therefore ...we must use *Induction*, true and legitimate induction, which is the very key of interpretation.[277]

The process of induction from experience to axiom and deduction from axiom to experiment is not the end of the matter. Rather, the results of experiment give a middle class of generalization which must be compared with negative instances and instances of degree before the highest possible generalization is reached. By setting out tables of presentation, exclusion and rejection and instances of degree, Bacon proposes to take the axioms resulting from experiment through the most stringent process of elimination possible. The induction, deduction, experiment, elimination process results in the most valid identification of reality possible in each particular case.

As explained in the "Preface" to his *Instauration*, these tables are to form a "Natural History". In the *Novum Organum* this history is defined as "a natural and experimental history". As "the foundation of all" it lists things of "the same nature" in "tables and arrangements of instances".[278] Bacon illustrates the matter with the case of heat. Here the first table lists instances of the "single form" of heat. A second lists instances where the "form" listed by the first is wanting. A third indicates the instances of different degrees of the form. In order to determine "the dignities or prerogatives of instance" (twenty-seven in all) there are comparisons and notations of classifications, quantities, and matters of motion. Experiments are described. Instances of use are enumerated, bodies are divided into "primary" and subordinate, etc.[279] Finally, once again, "instances of magic" are included to cover cases "wherein the material or efficient cause is scanty or small, as compared with the work and effect produced; so that even where they are common, they seem like miracles".[280] For Bacon, magic covers a myriad of stop-gaps. It is the explanation of that which, after the most careful of investigations, cannot be otherwise explained.

Bacon's method, as shown by his examples of heat in his *Novum Organum*, has its limitations.[281] It is over-laborious and extremely tedious. His method of reaching axioms (hypotheses) via induction is hardly defined. His "forms" are too indefinite and, most importantly, his generalities are left in verbal form rather than being reduced by mathematics to numbered quantities. Nevertheless, the fact that Bacon recognized that the nature of nature consisted of the "forms" which "underlie" it and that the world was a complex of bodies, each of its own nature and of distinct qualities, color, taste, smell, specific hardness, density, etc., laid out the direction science was to take. These "natures" which are of limited number are nature's alphabet.[282] Like letters of a word, "simple natures" are combined to form the things of nature or "compound bodies, as they are found in nature in its ordinary course".[283] To take white as an instance, Bacon first designates it as a specific form found wherever white is found and never found where white is absent.[284] He then moves on to demonstrate the necessity of experiment by indicating that to know the form of white is to know how to produce it.[285]

The analysis of the way these forms or bodies are combined into structures results in the discovery of the "particular and special *habits* of nature", whereas the "universal *laws*" of nature constitute forms.[286] Hence, it is by ascertaining the laws and habits of nature that nature is analyzed and known. This occurs first "by reasoning" and "by true induction with experiments to aid", and then second by a "comparison of other bodies, and a reduction to simple natures and their Forms, which meet and mix in the compound".[287] Recombining the freshly recognized simple natures and forms leads to the real end of the matter — the transformation of nature.

> This mood of operation (which looks to simple natures though in a compound body) proceeds from what in nature is constant and eternal and universal, and opens broad roads to human power, such as (in the present state of things) human thought can scarcely comprehend or anticipate.[288]

For Bacon the proper penetration into the heart of nature is primarily the business of the rational facilities. "We must pass from Vulcan to Minerva, if we intend to bring to light the true textures and configurations of bodies".[289] Thus, it is the mind which experimentally tests its axioms. One need not chose between experimentation and reasoning; rather one should seek the proper combination of the two.

> The men of experiment are like the ant; they only collect and use; the reasoners resemble spiders, who make cobwebs out of their own substance. But the bee takes the middle course; it gathers its material from the flowers of the garden and of the field, but transforms and digests it by a power of its own.[290]

"Therefore", Bacon goes on, "from a closer and purer league between the two facilities, the experimental and the rational (such as have never yet been made), much may be hoped".[291]

Thus, Bacon's *Novum Organum,* his "new logic", or "new method" by which he aims "to teach and instruct the understanding",[292] includes indications of an idealist-realist mutual modification. The fact that Bacon insists that we know nature as nature and can say, "Let the images and rays of natural objects meet in a point, as they do in the sense of vision",[293] indicates that nature, for Bacon, is more than inert quantifiables. Bacon proposes a mind-nature complementarity. In this proposal we may begin to see an awareness of "inter-subjectivity" on his part. Thus, he anticipates that his method

> . . . may in very truth dissect nature, and discover the virtues and actions of bodies, with their laws as determined in matter; so that this

science flows not merely from the nature of the mind, but also from the nature of things.[294]

Centuries ahead of his time, Bacon intended to "hand over to men their fortunes". For "now their understanding is emancipated and come as it were of age". As a result, "there cannot but follow an improvement in man's estates, and an enlargement of his power over nature".[295] Bacon ends his *Novum Organum* with a statement that deserves to be the very motto of science and technology seen from the point of the Judeo-Christian faith.

> For man by the fall fell at the same time from his state of innocency and from his dominion over creation. Both of these losses however can even in this life be in some part repaired; the former by religion and faith, the latter by arts and sciences. For creation was not by the curse made altogether and forever a rebel, but in virtue of that charter, "In the sweat of thy face shalt thou eat bread", it is now by various labours (not certainly by disputations or idle magical ceremonies, but by various labours) at length and in some measure subdued to the supplying of man with bread; that is, to the uses of human life.[296]

Bacon was far from being an experimental scientist. He lacked the correct experimental mood. He questioned the Copernican theory,[297] and doubted if observers such as Galileo and observations made with the microscope or telescope were to be relied upon. Similarly he rejected the reliability of measurements taken with rods or the astrolabe.[298] Like Galileo, interestingly enough, he had so false a theory of the tides that when he noted the interval of six hours between them, he denied it as a delusion.[299] Unlike Descartes who wrote in French and Galileo who wrote in Italian, Bacon mistrusted modern languages. He believed that they were incapable of preserving and transmitting a book through the ages. Consequently he sought to translate all his English works into Latin.[300] And finally he never fully rid himself of the attraction of the occult, the belief in magic, astrology and secret writings.[301]

Nevertheless he established the direction which science was to go, as he spelled out his dream for the future in his *New Atlantis*. His "Solomon's house", with its buildings and apparatus for invention, its galleries exhibiting every invention and the statues of their inventors is complemented on a practical level by the publication of inventors throughout the land and giving counsel on plagues and diseases. This, along with the marvelous works and prayers for illumination, reminds one of nothing so much as a university specializing in modern science, a modern Cambridge, perhaps, with its Cavendish Laboratory, chapels and History and Philosophy of Science Library and Museum,[302] or a Massachusetts Institute of Technology with magnificent facilities for research and engineering as well as a chapel for worship.

Hence, although we may not be able to regard Bacon as a great scientist as far as original discovery is concerned, his innovations were of revolutionary significance for modern science. The methodology he proclaimed includes (1) the systematic gathering of data for experimentation with nature, (2) the systematic recording of the results of his experiments, and (3) the formulation of new hypotheses, the basis of which is to be tested by experiment, and the results of which are then to be assessed by comparison to others. The scientist must be willing to change his presuppositions on the basis of his method. This was to be fundamental for the methodology of experimental science.

Yet in the end Bacon's vision fell short. His distrust of pure quantitative mathematical explanation of science along with his commitment to other tasks inhibited his development of modern experimental science as such. Certainly A. N. Whitehead is correct when he states that the real difference between Bacon, on the one hand, and Galileo and Newton, on the other, was Bacon's inadequate appreciation of the quantitative aspects of science,[303] the possible limitations of which we have indicated. At the same time, however, in that science quickly began to fulfill the Pythagorean dream of seeing the earth only as number and hence, "What it can't count, it can't see", indicates that Bacon was probably far more farsighted, about 350 years more farsighted, than he is generally given credit for. We are just now beginning to become aware of the devastation wrought on creation by reducing it to the quantification of mathematical formulae as such.[304]

[At this point the text from the hand of Prof. Harold Paul Nebelsick ends. He died on Easter Sunday 1989]

NOTES

1. Cf. *Interpreter's Dictionary of the Bible*, 5 vols. (Nashville: Abingdon, 1962), I, 728.
2. I am indebted to this sense of "the contingency of the world upon God" to T. F. Torrance, *The Ground and Grammar of Theology* (Charlottesville, Va.: University of Virginia, 1980), pp. 53 ff.
3. It is thus more than a coincidence that the Hebrew תהום, *tehom*, meaning "the deep" or "the abyss", is related to "Tiamat". Cf. Ps. 104:7-9, Pr. 8:27-29.
4. Jaki, *Science and Creation*, p. 99. Cf. Jaki's chapter, "The Omen of the Ziggurats", *ibid.*, pp. 85-99; pp. 140 ff., where he recounts the documentary evidence for God's creation in the Psalms; and pp. 147 ff., where he shows the difference between the Enuma Elish and the Genesis creation accounts.
5. R. Hooykaas, *Religion and the Rise of Modern Science* (Grand Rapids: Eerdmans, 1978), p. 8. Jaki points out that "Yahweh God is an exclusive source of effectiveness. He is not challenged or complemented by any force or principle. He is the sole and supreme Lord of all." Jaki, *Science and Creation*, p. 140.
6. Plato, *Timaeus*, 41A, 37C.
7. Aristotle, *The Generation of Animals*, LCL (London: Heinemann, 1943), II. i, 734a.
8. A. R. Peacocke, *Creation and the World of Science* (Oxford: Clarendon Press, 1979), pp. 205 f.
9. C. H. Dodd, *The Interpretation of the Fourth Gospel* (Cambridge: University Press, 1953), p. 277. Cf. A. R. Peacocke, *Science and the Christian Experiment* (London: Oxford University Press, 1971), pp. 158 ff.
10. Dodd, *Interpretation of the Fourth Gospel*, p. 277. For a thorough study of the *logos* in Greek thought, cf. Jaki's chapter, "The Labyrinths of the Lonely Logos", *Science and Creation*, pp. 102-132.
11. Hooykaas, *Religion*, p. 7.
12. Peacocke, *Science and the Christian Experiment*, p. 120. Peacocke's statement, "The cosmos has contingent and derived being whereas only God has necessary and underived being, that is, an existence not dependent on any other entity", spells out this contingency relationship in metaphysical terms which would seem quite correct if one is careful to emphasize the unqualified difference between the "being" of God and that of the cosmos. *Ibid.* Cf. Peacocke's chapter, "God and the cosmos", *ibid.*, pp. 120-139.
13. Hooykaas, *Religion*, p. 8.
14. Torrance, *Theological Science*, p. 66. As indicated, even Thomas Aquinas was well aware that the *aeternitas mundi* had to be rejected on the basis of biblical authority, though it could not be on the

grounds of his (Aristotelian) philosophy.
15 *Ibid.*, p. 65.
16 Whitehead, *Science and the Modern World*, pp. 12 ff. Whitehead, in fact, points out that science arose as an anti-rationalistic movement in reaction to the rationalism of the scholastics. *Ibid.*, p. 16.
17 *Ibid.*, p. 12.
18 For Thomas' argument against Aristotle's *aeternitas mundi*, Nebelsick, *Circles*, pp. 149-154. The coincidence between Whitehead's concept of God of "primordial and consequent natures" with the Aristotelian-Thomistic God of first and final causes is rather obvious. In both Thomas and Whitehead God is responsible for but interpenetrates nature to the extent that in the end differentiation is quite impossible. Cf. A. N. Whitehead, *Process and Reality*, (Cambridge: University Press, 1929), pp. 486 ff., and A. N. Whitehead, *Religion in the Making*, (Cambridge: University Press, 1926) pp. 69 f.,156. For an extended discussion on the matter, cf. "Whitehead's God", Nebelsick, *Science and Theology*, pp. 48-52.
19 Torrance, *Divine and Contingent Order*, (New York, Oxford University Press, 1981) p. 3.
20 Wilhelm Dilthey, *Einleitung in die Geisteswissenschaften*, 6th ed., *Gesammelte Schriften* I (Stuttgart, Teubner, 1966), p. 272.
21 Whitehead, *Science and the Modern World*, p. 12. Whitehead, of course, makes no secret of his lack of appreciation or understanding of the Reformation. *Ibid.*, pp. 12 ff. It should also be pointed out perhaps that the statement can reflect modern physics only if the "general principles" are of statistical nature covering a multitude of individual events rather than "every detailed occurrence".
22 *Ibid.*
23 Jaki, *Relevance of Physics*, p. 433.
24 *Ibid.*
25 Bonhoeffer, who was later executed by the Nazis, was, at the time he wrote the statement from his prison cell in Berlin-Tegel, reading C. F. von Weizsäcker's *Weltbild der Physik (World-view of Physics)* in which Weizsäcker discusses Laplace's cosmology.
26 For an extended and profound discussion of the relationship of God to the world cf., T. F.Torrance, "God and the Contingent Universe", *Divine and Contingent Order*, pp. 26-61.
27 Jaki, *Science and Creation*, p. 158.
28 Torrance, *Theological Science*, p. 61.
29 Cf. Nebelsick, *Circles*, pp.1-41 and Eduard Zeller, *Die Philosophie der Griechen in ihrer geschichtlichen Entwicklung*, I. Teil, 1. Abt. (Hildesheim: Olms, 1963), pp. 55, 86f. 236 f., *et al.*
30 Cf. Torrance's discussion of Bacon in this regard, *ibid.*, pp. 69-70. Bacon himself credits Democritus with disposing of the doctrine of final causes. Bacon, *On the Dignity and Advancement of Learning, The Works of Francis Bacon*, ed. Spedding, Ellis, Heath, 7 vols.(London: Longman, 1857-1858), IV, Book 3, p. 365.

31 Plato, *Timaeus*, 27 ff. Cf. F. W. J. von Schelling, *Von der Weltseele* (Hamburg: Perth, 1809), pp. 3 ff., 8, 17, 27, *et al.*

32 Paul Tillich, *Systematic Theology*, 3 vols. (Chicago: University of Chicago Press, 1951-1963k), III, 421. For a more sympathetic evaluation of panentheism as a metaphorical concept expressing the relationship of God and creation, cf. Peacocke, *Creation and the World of Science*, pp. 45, 141, 201, *et al.* Tillich's concept of panentheism, "All in God", is similar to that of Charles Hartshorne whose idea is a resuscitation of that of K. C. F. Krause's mid-nineteenth century attempt to reconcile theism and pantheism with the subjectivism of Kant and Fichte and the romantic absolutism of Schelling and Hegel. Cf. Nebelsick, *Theology and Science*, p. 53; and *Religion in Geschichte und Gegenwart*, 3rd ed., 7 vols. (Tübingen, C. B. Mohr, 1961), V, 36 ff.

33 Hooykaas, *Religion*, p. 12.

34 Jaki, *Relevance of Physics*, pp. 417 f., Cf. Pierre Duhem, *Le Système du Monde*, 2nd ed., 10 vols. (Paris: Hermann, 1954-1959), VI, 66.

35 Crombie, *Augustine to Galileo*, I, 76.

36 Günter Howe, *Die Christenheit im Atomzeitalter* (Stuttgart: Klett, 1970), p. 263. Translation mine.

37 Howe, *Mensch und Physik*, p. 18.

38 Jaki, *Science and Creation*, p. 140.

39 Karl Barth, "Work of Creation", Section 41, *Church Dogmatics* III/1 (Edinburgh: T. & T. Clark, 1958). Cf. Torrance, *Theological Science*, p. 68.

40 Jaki, *Science and Creation*, p. 140.

41 Barth, "Work of Creation". For a view of science from the standpoint of the doctrine of creation, cf. Peacocke, *Creation and the World of Science*, esp. pp. 50-85 and 255-359.

42 Martin Heidegger, cited by Howe, *Christenheit*, p. 197.

43 *Ibid.*

44 Lynn White, "The Historical Roots of our Ecological Crisis", *Science*, 155, No. 3767 (March, 1967), 1207.

45 ClausWestermann, *Genesis, Kapitel 1-11* (Neukirchen-Vluyn: Neukirchener, 1974), pp. 219 f.

46 Norbert Lohfink, *Unsere grossen Wörter* (Freiburg: Herder, 1977), p. 169, cf., *ibid.*, pp. 167 ff. Lohfink adds, "There is not the least supposition of a deep ditch between human and sub-human creatures". Rather, according to the scheme of the priestly texts, we have here to do with "a paradise-like, peaceful unity in which the characteristics of the image of God should realize itself for the sake of humankind". *Ibid.*, p. 169.

47 Hence our objection to Llynn White's thesis with reference to "the orthodox Christian arrogance over nature".

48 J. Robert Nelson, *Science and Our Troubled Conscience* (Philadelphia: Fortress, 1980), pp. 72 ff. Einstein's "God does not play dice" sums up his thought regarding the inner causal-rationality of nature. Alfred Einstein and Max Born, *Born-Einstein Letters* (London: Macmillan,

1971), p. 91. For a full exposition of Einstein's position over against Heisenberg and Bohr, cf. A. Einstein, B. Podolsky and N. Rosen, "Can Quantum-Mechanical Description of Physical Reality Be Considered Complete?", *Physical Review*, 47, 1935, pp. 777-780.

49 Butterfield, *Origins of Modern Science*, pp. 98 f. Cf. André Dumas, "The Ecological Crisis and the Doctrine of Creation", *The Ecumenical Review*, 1 (1975), 24 f. The same concept is seen in Roger Bacon and also in Francis Bacon, *Novum Organum*, tr. R. Ellis and James Spedding (London: Routledge, n.d.), I, cxxix; II, lii.

50 Bacon, *Novum Organum*, I, cxxx; II, lii.Bacon, *Novum Organum*, I, cxxx; II, lii.

51 Klaus Koch, "The Old Testament View of Nature", *Anticipation*, 25 (January, 1979), 49.

52 *Ibid.*

53 St. Augustine, *The Confessions*, Great Books of the Western World, Vol 18 (Chicago: Encyclopaedia Britannica, 1952), XI. 26-29, p. 95. Cf. St. Augustine, *The City of God*, Great Books of the Western World, Vol. 18 (Chicago: Encyclopaedia Britannica, 1952), XII. 13, p. 350 for Augustine's discussion of Plato's cyclic concept of history.

54 Peter Brunner, "Zur Auseinandersetzung zwischen antiken und christlichem Zeit- und Geschichtsverständnis bei Augustin", *Zeitschrift für Theologie und Kirche*, NF, 14. Jg.. (1933), pp. 1-25. Cf. Oscar Cullmann, *Christ and Time* (Philadelphia: Westminster, 1964), pp. 51 ff. Cullmann is directly dependent upon G. Delling, *Das Zeitverständnis des neuen Testaments* (Gütersloh: Bertelsmann, 1940).

55 Thorlief Boman, *Hebrew Thought Compared with Greek* (London: SCM, 1960). Cf. James Barr, *The Biblical Words for Time* (Napierville, Ill.: Allenson, 1962).

56 Boman, *Hebrew Thought*, p. 125. The original citation is from Aristotle, *Physics* I, IV. xiv, 223b. cited by Delling, *Zeitverständnis*, p. 148 and from him by Cullmann, *Christ and Time*, p. 52. P. H. Wicksteed and F. M. Cornford translate the statement, For time itself is conceived as 'coming round'". Aristotle, *Physics*, IV. xiv, 223 b.

57 *Ibid.*

58 *Ibid.*, IV. xiv, 223a.

59 *Ibid.*, IV. x, 218b.

60 *Ibid.*, IV. xiii, 222b.

61 Brunner, "Auseinandersetzung", p. 2.

62 *Ibid.*, p. 3.

63 Plato, *Timaeus*, 37, p. 620.

64 *Ibid.*, 38, p. 620.

65 Boman, *Hebrew Thought*, p. 128.

66 *Ibid.*

67 Plato, *Symposium*, *Dialogues of Plato*, II, 2211, p. 61.

68 Aristotle, *Physics* I, IV. xi, 219a.

69 *Ibid.*

70 *Ibid.* Cf. H. Bergson, *Time and Free Will*, tr. F. L. Pogson (London:

224 The Renaissance, the Reformation and the Rise of Science

Sonnenschein, 1910), Chapter II for a discussion of time as conceived in terms of space because of our space-dominated minds.
71 Boman, *Hebrew Thought,* p. 126.
72 Aristotle, *Physics* I, IV. xii, 22lb.
73 *Ibid.*
74 Boman, *Hebrew Thought,* p. 124.
75 Thus, it is important to recognize that the Genesis 1:31 pronouncement of creation as "good" was written several centuries after the story of the "Fall: (Gen 2:15 ff.).
76 Barth, *Church Dogmatics* III/1, Section 41.
77 Butterfield, *Origins of Modern Science,* p. 213.
78 Karl Barth, *Das Evangelium in der Gegenwart* (München, Chr.-Kaiser, 1935), p. 32.
79 John Calvin, *Institutes of the Christian Religion,* Library of Christian Classics XX, XXI, ed. John T. McNeill (Philadelphia: Westminster, 1960), Book III, Ch. X. 3-5.
80 *Ibid.,* Book, Ch. X. 2.
81 Jaki, *Science and Creation,* pp. 164-187.
82 Jaki, *Relevance of Physics,* p. 52.
83 Max Born, *Physics in My Generation* (London: English Universities, 1970)., p. 120.
84 Howe, *Mensch und Physik,* p. 105, and *Christenheit,* pp. 196 f.
85 Georg Picht, *Theologie und Kirche im 20. Jahrhundert* (Stuttgart: Ernst Klett, 1972), p. 19.
86 Plato, *Protagoras, Dialogues* I, 320-322, pp. 134 f.
87 The fact that even today in the Levant, the long fingernail on the little finger of one or both hands remains a badge of the "non-labourer" indicates that the problem has not been entirely overcome.
88 Calvin, *Institutes,* Bk. III, Ch. X.6 is a key to R. S. Wallace's *Calvin's Doctrine of the Christian Life* (Grand Rapids: Eerdmans, 1959). Hence, Wallace remarks with reference to Calvin's doctrine of work, "Not only does God accept our works as righteous, he actually rewards us for them as if they were worthy of such reward, though such reward proceeds not from our merit but from His own undeserved grace", p. 302.
89 Butterfield, *Origins of Modern Science,* p. 93.
90 *Ibid.,* p. 162.
91 Bernard de Fontenelle, Éloge de Monsieur Ozanam", *Oeuvres de Monsieur Fontenelle,* Vol. 5 (Paris: Brunet, 1758), p. 560.
92 Butterfield, *Origins of Modern Science,* p. 163.
93 Butterfield, *Origins of Modern Science,* p. 24. Cf. Nebelsick, "Deities of Plato and Aristotle", *Circles,* [typ. pp. 89-102].
94 Cf. Howe, *Christenheit,* p. 137.
95 Torrance, *Theological Science,* pp. 65 ff.
96 Reijer Hooykaas, "Humanisme, Science et Réforme", *Free University Quarterly V* (1958), 278. Cf. Reijer Hooykaas, "Science and Reformation", *The Evolution of Science* (New York: New American Library,

1963), pp. 258-290.

97 Martin Luther, *Tischreden* [entries from 1538-1539 recorded by Antony Lauterbach] (Eisleben: Gaubisch, 1566). In the entry for June 4, 1539, p. 582, the German text indicates that Luther understood the Copernican system when he spoke of the three movements involved saying, " . . . the first is the prime mover". It moves the whole firmament so fast and nimbly that in twenty-four hours it circles around in one complete motion several thousand mile lengths (*meilenwegs*) sent on its way perhaps by an angel. It is a wonder that such a large structure (*Gewebe*) and globe should circle around and move in so short a time".

Neither Luther nor Calvin accepted the Copernican cosmology which was initiated in their own time and both questioned the heliocentric theory—Luther in his *Table Talks*, as mentioned above, and Calvin, in a sermon delivered in 1556, in which he also takes those to task who would say that "the sun did not move but that it is the earth which shifts and turns itself". Nevertheless, neither was opposed to the development of science *per se*. Both, in fact, expressed the deepest admiration for astronomy and the astronomers. The French of Calvin's statement, "Ils idront que le soleil ne se bouge, et que c'est la Terre que se remue et qu'elle tourne" is given by Pierre Marcel, "Place et Thème du huitième sermon sur "Corinthiens 10", *Calvin & Copernic: La Legend ou les Faits? La Science et l'Astronomie chez Calvin* (*La Revue Réformée* [Mars 1980] No. 121-1980/1, Tome XXXI), pp. 15-20. In this issue of *La Revue Réformée* the controversy as to whether or not Calvin was anti-Copernican is thoroughly reviewed. Cf. also Robert White, "Calvin and Copernicus, The Problem Reconsidered", *Calvin Theological Journal* (Nov. 1980), Vol. 15, No. 2, pp. 233-243. A discussion of the sermon in which Calvin warns against those who pervert truth, and to which the statement about the sun and the earth is incidental, is found in the *Calvin Theological Journal* (Nov. 1980), cf. p. 237, fn. 14.

For the record, it must also be pointed out that Andrew White's remark that Calvin condemned the Copernican theory directly has no basis in fact. According to White, "Calvin took the lead, in his *Commentary on Genesis*, by condemning all who asserted that the earth is not at the centre of the universe. He clinched the matter by the usual reference to the first verse of the ninety-third Psalm, and when asked, 'Who will venture to place the authority of Copernicus above that of the Holy Spirit?'" Andrew Dickson White, *A History of the Warfare of Science with Theology in Christendom*, 2 vols. (London: Macmillan, 1897), I, 127. The statement which unfortunately has been repeated by both Bertrand Russell and latterly by the respected historian of science Thomas Kuhn is simply not to be found in Calvin's writings.

White's source for "Calvin's statement" regarding Copernicus is an undocumented statement in the preface of the 1886 edition of

226   The Renaissance, the Reformation and the Rise of Science

Frederic W. Farrer's Bampton Lectures of 1885 entitled *History of Interpretation* (Grand Rapids, Mich.: Baker, 1961), p. xviii. Farrer, who otherwise thinks highly of Calvin, naming him "the greatest exegete and theologian of the Reformation" explains that though he "explains away every passage which runs counter to his dogmatic presuppositions . . . he is one of the greatest interpreters of Scripture who ever lived" (*Ibid.*, pp. 342 f.), must simply have mis-credited the statement. Little wonder that Reijer Hooykaas says, "For fifteen years, I have pointed out in several periodicals concerned with the history of science that the 'quotation' from Calvin is imaginary and that Calvin never mentioned Copernicus; but the legend dies hard". Hooykaas, *Religion and the Rise of Modern Science*, p. 121. Hooykaas also indicates that "a parallel quotation allegedly from the Independent divine John Owen" is equally "spurious". *Ibid.* For Kuhn's statement cf. Thomas Kuhn, *The Copernican Revolution* (Cambridge: Harvard University Press, 1979), p. 192.

[98] Butterfield, *Origins of Modern Science*, pp. 180 ff.
[99] Günter Howe, *Gott und die Technik*, p. 69. The Royal Society of London which was dedicated to the promotion of science and chartered in 1662, was founded in 1660 and on November 28th of that year, its first journal was opened. A nucleus of persons interested in "natural philosophy" meet weekly in London as early as 1645. Cf. Hooykaas, *Religion and the Rise of Science*, pp. 135-148, for a discussion of "Puritanism and Science".
[100] Cf. Howe, *Christenheit*, p. 198.
[101] Max Weber, *The Protestant Ethic and the Spirit of Capitalism* (New York: Scribner, 1930).
[102] Butterfield, *Origins of Modern Science*, p. 165. Cf. Hooykaas, *Religion and the Rise of Science*, esp. the chapter, "Science and the Reformation", pp. 98 ff.
[103] John Spedding, ed., *Letters and Life of Francis Bacon*, 7 vols. London, 1861, Vol.I, iv.
[104] Bacon's father, Nicholas, was the Keeper of the Seal to Queen Elizabeth.
[105] Spedding, *Letters and Life*, I, 4, 8; III, 84 ff., 363 ff.; IV, 25 ff., 142 ff.; VII, 35 ff., 135, 285, 371 ff., 377 ff., 530 f.
[106] *Ibid.*, I, 2 f.
[107] Deason, p. 8.
[108] John Spedding, ed., *The Life and Acts of Matthew Parker*, 4 vols., Oxford 1821, vol 2, pp. 69ff.
[109] *Works* Vol XII, pp. 89
[110] *Letters*, Vol. 1, p. 85.
[111] Deason, p. 15.
[112] *Works* XII, p. 90.
[113] *Letters* I, p. 86.
[114] *Letters* III, p. 107.
[115] *Letters* I, pp. 185-188.

116 *Letters* I, p. 50.
117 *Letters* I, p. 5.
118 *Works* XI, p. 455.
119 *Letters* I, p. 81.
120 *Letters* I, p. 41.
121 *Works* VI, p. 408.
122 *Letters* I, p. 88.
123 *Letters* I, p. 101, 109.
124 *Letters* I, p. 92.
125 *Letters* III, p. 85 f.
126 *Works* XIII, pp. 83-93, 107-108.
127 *Works* XII, p, 160 f.
128 *Letters* I, p. 90.
129 *Letters* V, p. 90
130 *Letters* I, p. 90f.
131 *Letters* I, p. 91.
132 *Works* XIV, p. 89.
133 *Works* XII, p. 11.
134 *Refutation of Philosophies,* from Farrington in *The Philosophy of Francis Bacon,* p. 124.
135 *Works* VI, p. 28.
136 *Works* VI, pp. 33 f.
137 *Works* VI, p. 94.
138 *Works* VI, p. 65.
139 *Works* XII, p.113.
140 *Ref. Phil.* p. 128.
141 *Ibid.*
142 *Ibid.,* p. 131.
143 *Letters* III, p. 84 f.
144 *Ibid.*
145 Bethell, p. 40 f. He cites Jean Martin's 1550 translation of De Sebonde's *Natural Theology* of 1491, p. 25 in his work, *The Cultural Revolution of the Seventeenth Century,* citing Tillard's presentation of Raymond de Sebonde's work.
146 *Ibid.,* he cites Alexander Pope in his *Essay on Man,* Epistle I, Section VIII, 1.237.
147 *Ibid.* He refers to Arthur O. Lovejoy, *The Great Chain of Being,* Harvard Universtiy Press, 1936).
148 For speculations of Bacon's being responsible for Shakespeare's plays first made by Lord Donnely, as well as his supposed relationship with such cults and institutions as the Rosicrucians and Masons, cf. W.F.C. Wigston, *Bacon, Shakespeare and the Rosicrucians* (London: Redway, 1888) and Mrs. Henry Pott, *Francis Bacon and his Secret Society* (London: Banks, 1911). It must be admitted that Bacon followed the customs of his time more closely than most, perhaps, if not revealing all his inmost thought the fact that he utilized cipher in some of his writing adds to the mystery. Nevertheless, it would seem

possible that the "Baconisms" in Shakespeare are due to a coincidence of milieu and source material. Likewise, the similarity of the images of the *New Atlantis* and *De Sapienta Veterum* as well as Bacon's allusions to ancient occult and mythical materials and the teachings of the Rosicrucians and Masons may very well be explained on the basis of the Renaissance encyclopaedic tradition and Hermetic materials which Bacon also used as sources. The fact that we are often unfamiliar with said sources and milieu except through Bacon opens the possibility of making him responsible for much of the mystic and occult that had become common coin in the late Renaissance but later was largely suppressed in the age of science.

149 Francis Bacon, "Letter to Father Baranzane, Professor of Philosophy and Mathematics at Annecy" in *The Letters and Life of Francis Bacon*, 7 vols. edited by J. Spedding (London: Longman, Green, 1861-68), VII, 378.

150 Francis Bacon, "Letter to the Bishop of Winchester", *Letters and Life*, VII, 372.

151 *Ibid.*, pp. 371 ff. where Bacon outlines his purpose in devoting himself to the "Advancement of Learning".

152 Francis Bacon, *Valerius Terminus, Of the Interpretation of Nature, The Works of Francis Bacon*, 7 vols, ed. J. Spedding, R. L. Ellis and D. D. Heath, (London: Longman, 1857-58), III, 226.

153 *Ibid.*, p. 227.

154 *Ibid.*, p. 228.

155 Francis Bacon, "Preface", *Great Instauration* in *Works*, IV, 14.

156 Agrippa's work was translated and published in England, 1569-75.

157 V. F. Whitaker, *Francis Bacon's Intellectual Milieu* (Los Angeles University of California, 1962), p. 5.

158 Singer, *Giordano Bruno*, p. 182, n. 3.

159 *Ibid.*, p. 182.

160 Francis Bacon, *Novum Organum*, tr. R. Ellis and J. Spedding (London, n.d.), I, xlvi. The *Novum Organum* is also incorporated into *Works*, IV, 47 ff. The usual method of reference to book and aphorism is followed here.

161 Francis Bacon, *Of the Dignity and Advancement of Learning*, *Works*, IV, Book III, pp. 359-60. Cf. William Gilbert, (London, 1600).

162 Francis Bacon, *Redargutio Philosopharium*, *Works*, III, 571.

163 Bacon, *Advancement of Learning*, *Works*, III, Book I, p. 263.

164 Bacon, *Interpretation of Nature*, *Works*, III, 219.

165 For a much fuller treatment of the relationship of Francis Bacon to Hermeticism and science, cf. Charles Lenni, *The Classic Deities in Bacon* (Baltimore, 1933) and Walter Frost, *Bacon und die Naturphilosophie* (München, 1926).

166 R. L. Ellis, "Preface to the *Historia Densi et Rari* ", *Works*, II, 229 ff; Bacon, *Novum Organum*, I, cxvi. Cf. Whitaker, *Bacon*, p. 7.

167 Francis Bacon, *History of the Winds*, *Works*, V, 172.

168 Cf. Wigston, *Bacon*, esp. pp. 109 ff. and Pott, *Bacon*, esp. pp. 14 ff. The

champions of Baconian precedence for the Masonic order and the Rosicrucians seize upon this last reference.

169 Francis Bacon, *New Atlantis, Works*, III, 129 ff.
170 Spedding, *Letters and Life*, I, p. 4.
171 Bacon, *Advancement of Learning, Works*, III, 344.
172 Bacon's notes, "Comentarius solutus sive pandecta sive ancilla memoria", Spedding, *Letters and Life*, IV, p. 64. Cf. Francis Bacon, *Cogitata et Visa, Works*, III, 604.
173 Spedding, *Letters and Life*, IV, 143.
174 Francis Bacon, *The Wisdom of the Ancients written in Latine by the Right Honorable Sir Francis Bacon, Englished by Sir Arthur Gorges* (Edinburgh, Swinton, 1681), "The Preface".
175 *Ibid.*, p. 65.
176 *Ibid.*, p. 69.
177 *Ibid.*, p. 70.
178 *Ibid.*, p. 72.
179 *Ibid.*, p. 157.
180 *Ibid.*, pp. 157 f.
181 *Ibid.*, p. 158.
182 *Ibid.*
183 *Ibid.*, p. 160.
184 *Ibid.*
185 Cf. esp. Bacon, *Advancement of Learning, I in Works*, III, 264 ff.; *Interpretation of Nature*, Ch. 1 in *Works*, III, 217 ff.
186 Spedding, *Letters and Life*, III, 84.
187 Francis Bacon, "Epistle Dedicatory", *Great Instauration in Works*, IV, 11 f.
188 Bacon, *Interpretation of Nature*, Ch. 1 in *Works*, III, 218.
189 Bacon, "Preface", *Great Instauration* in *Works*, IV, 20.
190 Bacon, *Cogitata et Visa* in *Works*, III, 611.
191 Bacon, *Interpretation of Nature*, Ch. 1 in *Works*, III, 221 f.
192 Spedding, *Letters and Life*, III, 84 f.
193 *Ibid.*, pp. 85 f. "Generals." = generalizations.
194 Torrance, *Theological Science*, p. 69. For Torrance's discussion of Bacon, cf., *ibid.*, pp. 69-75.
195 Bacon, *Interpretation of Nature*, in *Works*, III, Ch. 1, p. 221. Cf. Bacon, *Novum Organum*, II, lii.
196 Bacon, *Interpretation of Nature*, in *Works* III, Ch. 1, p. 220.
197 *Ibid.*
198 Bacon, *Cogitata et Visa* in *Works*, III, 611.
199 Bacon, *Novum Organum*, I, cxxix.
200 Bacon, *Interpretation of Nature*, in *Works*, III, Ch. 1, p. 224.
201 *Ibid.*
202 Spedding, *Letters and Life*, I, 3.
203 *Ibid*, I, 4.
204 *Ibid.*
205 Bacon, *Advancement of Learning* in *Works*, III, Book I, p. 285.

206 Bacon, *Interpretation of Nature* in *Works*, III, Ch. 4, p. 227.
207 Kuno Fischer, *Francis Bacon and seine Nachfolger* (Leipzig: Brockhaus, 1875), pp. 18 f.
208 *Ibid.*, p. 14.
209 *Ibid.*, p. 15.
210 *Ibid.*
211 Bacon, *New Atlantis* in *Works*, III, 140 ff.; 157, et. al.
212 Ellis, "Preface" to *De principiis atque originibus* in *Works*, III, 74; cf. editor's note, *Works*, I, 564, n. 1 to *De Augmentis Scientiarium*, Liber Tertius; Bacon, *Advancement of Learning* in *Works*, IV, Book III, p. 359; *On Principles and Origins*, in *Works*, V, 476.
213 Bacon, *On Principles* in *Works*, V, 465. He calls him "*Magnus pentathlus*", and notes that he is "treated as childish by the vulgar".
214 Ellis, "General Preface to Bacon's Philosophical Works", in *Works*, I, 45; Bacon, *Sylva Sylvarum: or A Natural History* in *Works*, II, 381; Bacon, *Interpretation of Nature* in *Works* III, 228; Bacon, *Novum Organon* in *Works*, IV, pp. 58, 60; *Advancement of Learning* in *Works*, IV, 363, 365; Bacon, *Thoughts on the Nature of Things* in *Works*, V, 419; Bacon, *On Principles* in *Works*, V, 465; Bacon, *Description of the Intellectual Globe* in *Works*, V, 514.
215 Bacon, *Novum Organum*, II, viii.
216 Ellis, "General Preface", *Works* I, 45.
217 Bacon, "Epistle Dedicatory", *Great Instauration* in *Works*, IV, 11.
218 Bacon, *Advancement of Learning* in *Works* III, Book II, p. 351.
219 Bacon, "Epistle Dedicatory", *Great Instauration* in *Works*, IV,11 f.
220 *Ibid.*, p. 12.
221 Bacon, "Preface", *Great Instauration* in *Works*, IV, 14.
222 *Ibid.*, p. 15. The simile is used in *Interpretation of Nature*, vii, to condemn Plato and Aristotle in relation to the weightier pre-Socratics. Bacon, *Interpretation of Nature* in *Works*, III, 227.
223 Bacon, "Preface", *Great Instauration* in *Works*, IV, 17.
224 *Ibid.*, p. 18.
225 Bacon, *Advancement of Learning* in *Works*, IV, Book II, p. 298. Cf. Bacon, *Advancement of Learning* in *Works*, III, Book II, p. 333; Francis Bacon, *Wisdom of the Ancients* in *Works*, VI, xlii; 725 f.
226 Bacon, "Preface", *Great Instauration* in *Works*, IV, 18. Cf.T. F. Torrance, *Theological Science*, p. 71.
227 *Ibid.*, IV,17.
228 *Ibid.*, IV,19.
229 *Ibid.*, IV,18.
230 Bacon, *Advancement of Learning* in *Works*, III, Book II, p. 325.
231 Bacon, "Preface", *Novum Organon* in *Works*, IV, 39.
232 *Ibid.*, IV, 40.
233 *Ibid.*, IV, 42.
234 Bacon, *New Atlantis* in *Works*, III, 156.
235 Bacon, *Novum Organum*, tr. R. L. Ellis and J. Spedding (London: Routledge, n.d.) I, xxxix-xliv. Cf. Bacon, *Interpretation of Nature* in *The*

Works of Francis Bacon, eds. J. Spedding, R. L. Ellis, and D. D. Heath, 7 vols.(London, Longman, 1857-58), III, Ch. 16, p. 245. Much as Roger Bacon may serve as a precursor of Francis, as R. L. Ellis points out, it is highly unlikely that Francis Bacon encountered Roger's "scientific writings". Ellis, "Preface", *Novum Organum*, p. 26; 69, n. 8.

236 Bacon, *Novum Organum*, I, xi, xii, xx, xlv, p. 72, n. 14. Cf. Bacon, *Advancement of Learning* in *Works*, IV, Book III, p. 348. Bacon, *Novum Organum*, I, xlv. Here "draco" means, as usual in the older astronomy, the projected circle of the moon's orbit.

237 Francis Bacon, *Preparative Towards a Natural and Experimental History* in *Works*, IV, Ch. IV, pp. 255 f.

238 Bacon, *Novum Organum*, I, xcv.

239 Bacon, "Preface", *Instauration* in *Works*, IV, 14.

240 *Ibid.*, p. 18.

241 Bacon, *Interpretation of Nature*, in *Works*, III, Ch. 1, p. 222.

242 Bacon, *Preparative* in *Works*, IV, Book II, p. 254.

243 *Ibid.*, Book VII, p. 259.

244 *Ibid.*

245 *Ibid.*, Book IX, pp. 261 f.

246 Francis Bacon, *Catalogue of Particular Histories by Titles* in *Works*, IV, 270.

247 A. E. Taylor, *Francis Bacon, British Academy Annual Master-Mind Lecture, read December 11, 1926* (London, 1926), p. 10. Levi's book is entitled, *Il pensiero di Francesco Bacone,*

248 The innumeration of "the idols", above from Book I of the *Novum Organum* is done for the sake of continuity of presentation.

249 Bacon, *Novum Organum*, I, liv.

250 Bacon, *Advancement of Learning* in *Works*, IV, Book V, p. 407.

251 Bacon, *Novum Organum* I, vi.

252 Whitaker, *Bacon*, p. 16; Bacon, *Advancement of Learning* in *Works* IV, Book III, p. 365. Ramus' two works in this direction are : *Dialecticae Institutiones* (1543) and *Aristoteliae Animadversiones* (1543). This language however may very well come from L. Valla and Rodolf Agricola and not P. Ramus. See T. F. Torrance, *Theological Science*, p. 70.

253 Bacon, *Advancement of Learning I* in *Works* III, 288.

254 Bacon, *Novum Organum*, I, viii.

255 *Ibid.*, I, xix.

256 Bacon, "Letter to Domine Baranzane", dated June, 1622 in Spedding, *Letters and Life*, VII, 378.

257 Bacon, *Cogitata et visa* in *Works*, III, 611.

258 Bacon, *Novum Organum* I, xxiv.

259 *Ibid.*, I, xxvi; cf. Bacon, *Interpretation of Nature* in *Works*, III, Ch. 17, pp. 245 f.

260 Bacon, *Novum Organum*, I, xxvi.

261 *Ibid.*, I, iii.

262 *Ibid.*, I, iv.

232   The Renaissance, the Reformation and the Rise of Science

263 Bacon, *Novum Organum*, in *Works*, IV, 37. Hence, Torrance is quite correct in pointing out that Karl Popper, who agrees that "there is no such things as a logical method of having new ideas" fails to appreciate that Bacon had long since made the point. Karl Popper, *The Logic of Scientific Discovery* (London: Hutchinson, 1959), p. 32, cited by Torrance, *Theological Science*, p. 70, n. 6.
264 Bacon, "Letter to Domine Baranzane", dated June, 1622, in Spedding, *Letters and Life*, VII, 376 f.
265 Bacon, *Novum Organum*, I, xxxi.
266 *Ibid.*, II, i.
267 *Ibid.*, II, ii.
268 Bacon, *Interpretation of Nature* in *Works*, III, Ch. 7, p. 222.
269 Bacon, *Novum Organum*, II, iii.
270 *Ibid.*, II, iv.
271 Bacon, *Novum Organum*, II, iv.
272 *Ibid.*, II, vi.
273 *Ibid.*, II, vii.
274 *Ibid.*, II, viii.
275 Bacon's inconsistency at this point can be seen from his statement to Domine Baranzane, "Be not troubled about the Metaphysics. When true Physics have been discovered, there will be no Metaphysics. Beyond the true Physics is divinity only".(*ultra quam nihil praeter divina*). Spedding, *Letters and Life*, VII, 377.
276 Bacon, *Novum Organum*, II, ix. "Natural Magic" which with alchemy and astrology Bacon calls, "Sciences of the imagination". Thus he says: "The sciences of Imagination pretendeth to call and reduce natural philosophy from a variety of speculations to the magnitude of works". Bacon, *Advancement of Learning* in *Works*, III, Book I, p. 289.
277 Bacon, *Novum Organum*, II, x.
278 *Ibid.*
279 *Ibid.*, II, xi-l.
280 *Ibid.*, II, li.
281 *Ibid.*, II, xi-xx.
282 Bacon, *Advancement of Learning* in *Works*, IV, Book III, p. 361. Cf. Francis Bacon, *The Alphabet of Nature* in *Works*, V, pp. 208 ff.
283 Bacon, *Novum Organum*, II, v.
284 Bacon, *Advancement of Learning* in *Works*, IV, Book III, p. 361. Cf. Bacon's further explanation of "whiteness" in *Interpretation of Nature* in *Works* III, Ch. 11, pp. 236 f.
285 Bacon, *Advancement of Learning* in *Works* IV, Book III, p. 361.
286 Bacon, *Novum Organum* II, v.
287 *Ibid.*, II, vii.
288 *Ibid.*, II, v.
289 *Ibid.*, II, vii. Cf. Bacon, *Advancement of Learning* in *Works*, III, Book II, p. 325; "To the King", *Advancement of Learning* in *Works* IV, Book II, p. 287.
290 Bacon, *Novum Organum*, I, xcv. In his *Advancement of Learning*, in

Works, III, Book I, p. 285, it is the schoolmen who "out of no great quantity of matter and infinite agitation of wit spin out unto us those laborious webs of learning which are extant in their books".
291. Bacon, *Novum Organum*, I, xcv.
292. *Ibid.*, II, lii.
293. Bacon, "Epistle Dedicatory", *Great Instauration* in *Works*, IV, p. 19.
294. Bacon, *Novum Organum*, II, lii.
295. *Ibid.*, Italics added.
296. *Ibid.*
297. Bacon, *Advancement of Learning* In *Works* III, Book II, p. 365; and in *Works* IV, Book III, p. 348.
298. Bacon, *Novum Organum*, II, xxxix.
299. Ellis, "Preface", *De fluxu et refluxu maris* in *Works*, III, 44; Bacon, *Novum Organum* II, xlvi.
300. Spedding, *Letters and Life*, VII, 434.
301. Bacon, *New Atlantis* in *Works*, III, 138. Cf. letter dated 1517 reflecting on "some astrologers in Italy" should quit their foolishness and "tell us that when all the planets except the moon are beyond the line in the other hemisphere for six months together, we must needs have a cold winter", Spedding, *Letters and Life*, VI, "Preface", p. xvi.
302. Bacon, *New Atlantis* in *Works*, II, 157 ff.
303. Alfred North Whitehead, *Science and the Modern World*, p. 45. Also Alfred North Whitehead may well be correct in pointing out that the process of induction is more complicated than Bacon anticipated. *Ibid.*, p.43.
304. Cf. especially the writings of C. F. von Weizsäcker and A.M. Klaus Müller both of whom point out that the purely quantifying endeavors of modern science tend to blind it to reality as a whole and, hence, while intending to preserve it in many cases, is the devastation of that which it would safeguard. As Weizsäcker points out with regard to "the higher organized organic systems", "if physics sets out to prove itself as correct in the last detail on the concrete basis of the organisms as such, these organisms must be destroyed", Müller, *Präparierte Zeit*, p. 415, citing C. F. von Weizsäcker's discussion with physicists and philosophers in Cambridge in *Quantum Theory and Beyond*, Theodore Bastin, ed. (Cambridge: University Press, 1971), p. 328. Or, as Müller puts it, Anthropomorphically interpreted physical experiment means 'Geichtschaltung' (the elimination of opposing realities) and, thereby, applies to subjects having personal centers, *eo ipso inhuman* resulting in a type of inhumanity which at the same time is hidden from the specific-scientific consciousness through alienation. *Ibid.*, p. 622.

Or, as M. Klaus Müller puts it, anthropomorphically interpreted physical experiment means 'Gleichschaltung' (the elimination of opposing realities), and thereby applies to subjects having personal centers. It is *eo ipso* inhuman, resulting in a type of inhumanity which at the same time is hidden from the specific-scientific consciousness through alienation. *Ibid.*, p. 622.

EPILOGUE

I first met Harold Nebelsick in October 1953 when he came to Scotland and was enrolled as a Ph.D. candidate in the University of Edinburgh. He had already majored in Philosophy at the University of Nebraska and graduated Bachelor of Divinity at San Francisco Theological Seminary *cum laude*. He had been drawn to New College, Edinburgh, by the reputation of John Baillie. That was the during the heyday of high ecumenical dialogue that followed the World Conference of Faith and Order at Lund in 1952, in which John Baillie and I were heavily involved. Harold was set to work on the Ecclesiology of Charles Gore which he completed in 1956, gaining a distinguished Ph.D. His years of research at New College were fruitfully interlaced with further studies on the Continent, at Göttingen, Berlin, Basel and Paris. Soon his main interest switched from philosophical theology to Christian Dogmatics, and he devoured the *Church Dogmatics* of Karl Barth. At the same time he began to develop his life-long interest in the interrelations between theology and natural science, not least in their underlying epistemological structures.

The second period in Dr Nebelsick's life found him first in Berlin and then in Beirut, during which his remarkable combination of ecumenical, dogmatic and scientific theology began to take their distinctive shape. In 1956 under the auspices of the Commission on Ecumenical Mission and relations of the Presbyterian Church, UPSA, Harold and Melissa were sent to Berlin and assigned as Fraternal Workers to the Church of Berlin-Brandenburg, where he served as one of the Pastors to the Neu-Westend Gemeinde until 1961. The next two years he spent back in the USA, in post-Doctoral Studies at Princeton Theological Seminary on Ecumenical Ecclesiology and as Consultant for Studies for the Commission on Ecumenical Mission and Relations of the United Presbyterian Church in New York. Then in 1963 he was appointed to the Near East School of Theology, Beirut, where he served as Professor of Systematic Theology, and from 1966 also as Vice-President, until 1968. Several of Harold's early essays were published there in the Near East School of Theology Quarterly, one of which significantly bore the title "Theology as Science and Proclamation".

The third period in Dr Nebelsick's career began with his appointment in 1968 as Professor of Doctrinal Theology, at Louisville Theological Seminary, when he set about thinking out in a more rigorous way the scientific nature and epistemological structure of Christian theology, and working out the mutual relations between theology and science. For this purpose he spent some time in Göttingen in the academic year 1975-76 making himself familiar with what had been done in the famous Dialogue carried on there after the war under the guidance of Günter Howe and Carl F. von Weizsäcker between physicists and theologians. And then in order to equip himself still further in this interdisciplinary field, he spent many months in Edinburgh from 1979 to 1981 working on the relations between theology and astronomy and theology and physics. These researches began to bear fruit in 1981 when he published *Theology and Science in Mutual Modification*, which was volume 2 in the series *Theology and Scientific Culture*. In 1981 he was made a Member of the Center of Theological Inquiry, in Princeton, which gave him the opportunity to continue research and writing in this field, resulting in a second volume entitled *Circles of God, Theology and Science from the Greeks to Copernicus*, 1985, contributed to the new series *Theology and Science at the Frontiers of Knowledge*. The completion of a third volume *The Renaissance, The Reformation and the Rise of Science*, designed to form a trilogy with the others, had to be delayed owing to work in preparing the vast *Who's Who in the Theology of Science* commissioned by Sir John Templeton.

Harold Nebelsick's sudden death on Easter Sunday, March 26, 1989 was a devastating blow to his dear wife,

Melissa, and his family, Louis Daniel, Mary Cevilla and James Henry. He owed an immense debt to Melissa who inspired, sustained, and helped him in all his research and writing. His unusually brilliant daughter and sons were a great joy to him, as was Paul Matheny his son-in-law. Harold Nebelsick was one of the finest and most upright people I have been given to know, who did not shrink from the cost of facing up squarely to the truth. He was also a man of abounding humour and compassion, to whom not a few of our ablest contemporaries, as well as a host of students, looked as a wise father in God and an ever-supportive friend. All his ministry was biblically based and evangelically directed. It was governed by a profound sense of the sovereign majesty of the living God and the presence of the risen Lord. One of his earliest essays in Beirut was entitled "Christ is Risen" - how fitting it was for such a follower of our Lord to pass into his immediate presence on Easter Day!

Harold's death was also an immense loss to the cause of Christian theology and scientific culture, not least to us his colleagues in the Center of Theological Inquiry. He was a Reformed theologian of great stature who was yet ecumenically sympathetic to other traditions where serious theological dialogue was concerned. He set for himself high standards of rigour and integrity which few others could match, but always cherished an open-minded understanding of what God has revealed of himself through the Holy Scriptures and revealed of his creation through the discoveries of natural science. He consistently declined to hold them apart, but sought to find adequate ways of thinking them together. In drawing out analogies between the epistemological structures of theological and natural science, he was indebted to some of the greatest minds of our time, such as Karl Barth, Niels Bohr and Michael Polanyi. He helped us considerably in clarifying the tangled connections between theology and natural science that have characterised western thought since Galileo Galilei and Isaac Newton. But he went further in offering us the results of his deep probing into the mutual modification between theological and scientific inquiry through the ages, not least in the light of the cosmological revolution that has taken place in our times.

Harold has left a signal mark, not least through the Center of Theological Inquiry, upon the exciting enterprise of theological and scientific integration upon which we have embarked. He was a driving force in the CTI Consultations held between theologians and scientists in 1989, in Oxford, Heidelberg, Burlingame, and Princeton. He chaired those consultations and guided our discussions with such insight and understanding, and irrepressible humour, that he stamped his authority upon them all. We will miss him greatly when we come to gather the fruit of all we have done in a final session. It was Harold's hope, as it has been mine, that through the Center of Theological Inquiry there could be established in the foundations of knowledge a reconciling interchange between theology and science in our understanding of God and the created universe, which would go far to affect the whole face of human life and culture and heal it of his divisions. This would give the Center of Theological Inquiry a unique place in the academic and scientific world.